A Primer of Conservation Genetics

The biological diversity of our planet is rapidly being depleted due to direct and indirect consequences of human activities. As the size of animal and plant populations decreases, loss of genetic diversity reduces their ability to adapt to changes in the environment, with inbreeding depression an inevitable consequence for many species. This concise, entry-level text provides an introduction to the role of genetics in conservation and presents the essentials of the discipline. Topics covered include:

- loss of genetic diversity in small populations
- inbreeding and loss of fitness
- resolution of taxonomic uncertainties
- genetic management of threatened species
- contributions of molecular genetics to conservation

The authors assume only a basic knowledge of Mendelian genetics and simple statistics, making the book accessible to those with a limited background in these areas. Connections between conservation genetics and the wider field of conservation biology are interwoven throughout the book.

The text is presented in an easy-to-follow format, with main points and terms clearly highlighted. Worked examples are provided throughout to help illustrate key equations. A glossary and suggestions for further reading provide additional support for the reader and many beautiful pen-and-ink portraits of endangered species help bring the material to life.

Written for short, introductory-level courses in genetics, conservation genetics and conservation biology, this book will also be suitable for practising conservation biologists, zoo biologists and wildlife managers needing a brief, accessible account of the significance of genetics to conservation.

DICK FRANKHAM was employed in the Department of Biological Sciences at Macquarie University, Sydney for 31 years and was Hrdy Visiting Professor at Harvard University for spring semester 2004. He holds honorary professorial appointments at Macquarie University, James Cook University and the Australian Museum.

JON BALLOU is Head of the Department of Conservation Biology at the Smithsonian Institution's National Zoological Park.

DAVID BRISCOE is Associate Professor at the Key Centre for Biodiversity and Bioresources, Department of Biological Sciences, Macquarie University, Sydney.

A Primer of
Conservation Genetics

Richard Frankham
Macquarie University, Sydney

Jonathan D. Ballou
Smithsonian Institution, Washington, DC

David A. Briscoe
Macquarie University, Sydney

Line drawings by
Karina H. McInnes
Melbourne

CAMBRIDGE
UNIVERSITY PRESS

CAMBRIDGE UNIVERSITY PRESS
Cambridge, New York, Melbourne, Madrid, Cape Town, Singapore, São Paulo, Delhi

Cambridge University Press
The Edinburgh Building, Cambridge CB2 8RU, UK

Published in the United States of America by Cambridge University Press, New York

www.cambridge.org
Information on this title: www.cambridge.org/9780521538275

First published 2004
Reprinted 2005, 2007

A catalogue record for this publication is available from the British Library

Library of Congress Cataloguing in Publication data
Frankham, Richard, 1942–
A primer of conservation genetics / Richard Frankham.
       p.   cm.
Includes bibliographical references and index.
ISBN 0 521 83110 5 (hardback) – ISBN 0 521 53827 0 (pbk.)
1. Ecological genetics.   2. Genetics.   3. Conservation biology.   4. Biological
diversity.   5. Evolutionary genetics.   I. Title.
QH456.F73   2003
576.5′8 – dc21   2003055125

ISBN 978-0-521-83110-9 hardback
ISBN 978-0-521-53827-5 paperback

Transferred to digital printing 2009

Contents

Preface

The World Conservation Union (IUCN), the primary international conservation body, recognizes the crucial need to conserve genetic diversity as one of the three fundamental levels of biodiversity. This book provides a brief introduction to the concepts required for understanding the importance of genetic factors in species extinctions and the means for alleviating them.

Conservation genetics encompasses the following activities:

- genetic management of small populations to retain genetic diversity and minimize inbreeding
- resolution of taxonomic uncertainties and delineation of management units
- the use of molecular genetic analyses in forensics and in improving our understanding of species' biology.

Conservation genetics is the use of genetics to aid the conservation of populations or species

Purpose of the book

We have endeavoured to make *A Primer of Conservation Genetics* as comprehensible as possible to a broad range of readers. It is suitable for those undertaking introductory genetics courses at university, for students undertaking conservation biology courses and even for motivated first-year biology students who have completed lectures on basic Mendelian genetics and introductory population genetics (allele frequencies and Hardy–Weinberg equilibrium). Conservation professionals with little genetics background wishing for a brief authoritative introduction to conservation genetics should find it understandable. These include wildlife biologists and ecologists, zoo staff undertaking captive breeding programs, planners and managers of national parks, water catchments and local government areas, foresters and farmers. This book provides a shorter, more basic entry into the subject than our *Introduction to Conservation Genetics*.

This book is intended to provide a brief accessible introduction to the general principles of conservation genetics

We have placed emphasis on general principles, rather than on detailed experimental procedures, as the latter can be found in specialist books, journals and conference proceedings. We have assumed a basic knowledge of Mendelian genetics and simple statistics. Conservation genetics is a quantitative discipline as its strength lies in its predictions. The book includes a selection of important equations, but we have restricted use of mathematics to simple algebra to make it understandable to a wide audience.

Mastery of this discipline comes through active participation in problem-solving, rather than passive absorption of facts. Consequently, worked **Examples** are given within the text for most equations presented. Many additional problems with answers provided can be found in our *Introduction to Conservation Genetics*.

Worked examples are provided

Due to the length constraints, references are not given in the text, but each chapter has **Suggestions for further reading**. Those wishing

Suggestions for further reading are provided

for detailed references supporting the assertions for particular topics will find them in our *Introduction to Conservation Genetics*.

Feedback, constructive criticism and suggestions will be appreciated (email: rfrankha@els.mq.edu.au).

We will maintain a web site to post updated information, corrections, etc. (http://consgen.mq.edu.au). On this site, choose the 'Primer' option.

Take-home messages

1. The biological diversity of the planet is rapidly being depleted due to direct and indirect consequences of human activities (habitat destruction and fragmentation, over-exploitation, pollution and movement of species into new locations). These reduce population sizes to the point where additional stochastic (chance) events (demographic, environmental, genetic and catastrophic) drive them towards extinction.

2. Genetic concerns in conservation biology arise from the deleterious effects of small population size and from population fragmentation in threatened species.

3. The major genetic concerns are loss of genetic diversity, the deleterious impacts of inbreeding on reproduction and survival, chance effects overriding natural selection and genetic adaptation to captivity.

4. In addition, molecular genetic analyses contribute to conservation by aiding the detection of illegal hunting and trade, by resolving taxonomic uncertainties and by providing essential information on little-known aspects of species biology.

5. Inbreeding and loss of genetic diversity are inevitable in all small closed populations and threatened species have, by definition, small and/or declining populations.

6. Loss of genetic diversity reduces the ability of populations to adapt in response to environmental change (evolutionary potential). Quantitative genetic variation for reproductive fitness is the primary component of genetic diversity involved.

7. Inbreeding has deleterious effects on reproduction and survival (inbreeding depression) in almost every naturally outbreeding species that has been adequately investigated.

8. Genetic factors generally contribute to extinction risk, sometimes having major impacts on persistence.

9. Inbreeding and loss of genetic diversity depend on the genetically effective population size (N_e), rather than on the census size (N).

10. The effective population size is generally much less than the census size in unmanaged populations, often only one-tenth.

11. Effective population sizes much greater than 50 ($N > 500$) are required to avoid inbreeding depression and $N_e = 500$–5000 ($N = 5000$–$50\,000$) are required to retain evolutionary potential. Many wild and captive populations are too small to avoid inbreeding depression and loss of genetic diversity in the medium term.

12. The objective of genetic management is to preserve threatened species as dynamic entities capable of adapting to environmental change.

13. Ignoring genetic issues in the management of threatened species will often lead to sub-optimal management and in some cases to disastrous decisions.

14. The first step in genetic management of a threatened species is to resolve any taxonomic uncertainties and to delineate management units within species. Genetic analyses can aid in resolving these issues.

15. Genetic management of threatened species in nature is in its infancy.

16. The greatest unmet challenge in conservation genetics is to manage fragmented populations to minimize inbreeding depression and loss of genetic diversity. Translocations among isolated fragments or creation of corridors for gene flow are required to minimize extinction risks, but they are being implemented in very few cases. Care must be taken to avoid mixing of different species, sub-species or populations adapted to different environments, as such outbreeding may have deleterious effects on reproduction and survival.

17. Genetic factors represent only one component of extinction risk. The combined impacts of all 'non-genetic' and genetic threats faced by populations can be assessed using population viability analysis (PVA). PVA is also used to evaluate alternative management options to recover threatened species, and as a research tool.

18. Captive breeding provides a means for conserving species that are incapable of surviving in their natural habitats. Captive populations of threatened species are typically managed to retain 90% of their genetic diversity for 100 years, using minimization of kinship. Captive populations may be used to provide individuals for reintroduction into the wild.

19. Genetic deterioration in captivity resulting from inbreeding depression, loss of genetic diversity and genetic adaptation to captivity reduces the probability of successfully reintroducing species to the wild.

Acknowledgments

The support of our home institutions is gratefully acknowledged. They have made it possible for us to be involved in researching the field and writing this book. The research work by RF and DAB was made possible by Australian Research Council and Macquarie University research grants. JDB gratefully acknowledges the Smithsonian National Zoological Park for many years of support. We are grateful to Barry Brook, Matthew Crowther, Vicky Currie, Kerry Devine, Mark Eldridge, Polly Hunter, Leong Lim, Annette Lindsay, Edwin Lowe, Julian O'Grady and to two anonymous reviewers for comments on drafts. We are grateful to Sue Haig and colleagues for trialling a draft of the book with the Applied Conservation Genetics course at the National Conservation Training Center in the USA in 2002 and to their students for feedback. We have not followed all of the suggestions from the reviewers. Any errors and omission that remain are ours.

We are indebted to Karina McInnes whose elegant drawings add immeasurably to our words. Michael Mahoney kindly provided a photograph of the corroboree frog for the cover, Claudio Ciofu provided the microsatellite traces for the Chapter 2 frontispiece, J. Howard and B. Pukazhenthi provided the sperm photograph in Box 7.1 and Nate Flesness, Oliver Ryder and Rod Peakall kindly provided information.

Alan Crowden from Cambridge University Press provided encouragement, advice and assistance during the writing of the book and Maria Murphy, Carol Miller and Anna Hodson facilitated the path to publication.

This book could not have been completed without the continued support and forbearance of our wives Annette Lindsay, Vanessa Ballou and Helen Briscoe, and our families.

Chapter 1

Introduction

Endangered species typically decline due to habitat loss, over exploitation, introduced species and pollution. At small population sizes additional random factors (demographic, environmental, genetic and catastrophic) increase their risk of extinction. Conservation genetics is the use of genetic theory and techniques to reduce the risk of extinction in threatened species

Terms

Biodiversity, bioresources, catastrophes, demographic stochasticity, ecosystem services, endangered, environmental stochasticity, evolutionary potential, extinction vortex, forensics, genetic diversity, genetic drift, genetic stochasticity, inbreeding, inbreeding depression, purging, speciation, stochastic, threatened, vulnerable

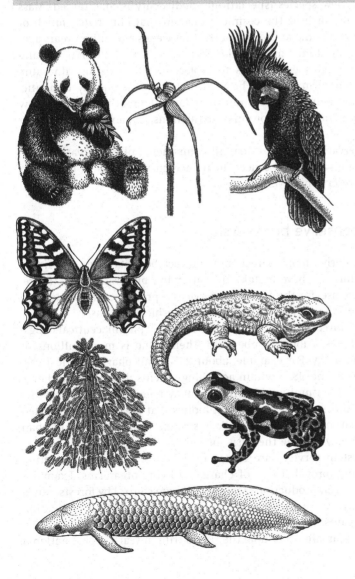

Selection of threatened species: Clockwise: panda (China), an Australian orchid, palm cockatoo (Australia), tuatara (New Zealand), poison arrow frog (South America), lungfish (Australia), Wollemi pine (Australia) and Corsican swallow-tail butterfly.

The 'sixth extinction'

The biological diversity of the planet is being depleted rapidly as a consequence of human actions

Biodiversity is the variety of ecosystems, species, populations within species, and genetic diversity among and within these populations. The biological diversity of the planet is rapidly depleting as direct and indirect consequences of human activities. An unknown but large number of species are already extinct, while many others have reduced population sizes that put them at risk. Many species now require human intervention to ensure their survival.

The scale of the problem is enormous and has been called the 'sixth extinction', as its magnitude compares with that of the other five mass extinctions revealed in the geological record. Extinction is a natural part of the evolutionary process, species typically persisting for ~5–10 million years. When extinctions are balanced by the origin of new species (**speciation**), biodiversity is maintained. Mass extinctions, such as the cosmic cataclysm that eliminated much of the flora and fauna at the end of the Cretaceous, 65 million years ago, are different. It took many millions of years for proliferation of mammals and angiosperm plants to replace the pre-existing dinosaurs and gymnosperm plants. The sixth extinction is equally dramatic. Species are being lost at a rate that far outruns the origin of new species and, unlike previous mass extinctions, is mainly due to human activities.

Conservation genetics, like all components of conservation biology, is motivated by the need to reduce current rates of extinction and to preserve biodiversity.

Why conserve biodiversity?

Four justifications for maintaining biodiversity are: economic value of bioresources; ecosystem services; aesthetic value; and rights of living organisms to exist

Humans derive many direct and indirect benefits from the living world. Thus, we have a stake in conserving biodiversity for the resources we use, for the ecosystem services it provides, for the pleasure we derive from living organisms and for ethical reasons.

Bioresources include all of our food, many pharmaceutical drugs, natural fibres, rubber, timber, etc. Their value is many billions of dollars annually. For example, about 25% of all pharmaceutical prescriptions in the USA contain active ingredients derived from plants. Further, the natural world contains many potentially useful new resources. Ants synthesize novel antibiotics that are being investigated for human medicine, spider silk is stronger weight-for-weight than steel and may provide the basis for light high-tensile fibres, etc.

Ecosystem services are essential biological functions benefiting humankind, provided free of charge by living organisms. Examples include oxygen production by plants, climate control by forests, nutrient cycling, water purification, natural pest control, and pollination of crop plants. In 1997, these services were valued at US$33 trillion ($10^{12}$) per year, almost double the US$18 trillion yearly global national product.

Many humans derive pleasure (**aesthetic value**) from living organisms, expressed in growing ornamental plants, keeping pets, visiting zoos, ecotourism and viewing wildlife documentaries. This translates into direct economic value. For example, koalas are estimated to contribute US$750 million annually to the Australian tourism industry.

The **ethical** justifications for conserving biodiversity are simply that our species does not have the right to drive others to extinction, parallel to abhorrence of genocide among human populations.

Endangered and extinct species

Recorded extinctions

Extinctions recorded since 1600 for different groups of animal and plants on islands and mainlands are given in Table 1.1. While over 700 extinctions have been recorded, the proportions of species that have become extinct are small, being only 1–2% in mammals and birds. However, the pattern of extinctions is concerning, as the rate of extinction has generally increased with time (Fig. 1.1) and many species are now threatened. Further, many extinctions must have occurred

Over 800 extinctions have been documented since records began in 1600, the majority being of island species

Table 1.1 | Recorded extinctions, 1600 to present, for mainland and island species worldwide

Taxon	Total	Percentage of taxon extinct	Percentage of extinctions on islands
Mammals	85	2.1	60
Birds	113	1.3	81
Reptiles	21	0.3	91
Amphibians	2	0.05	0
Fish	23	0.1	4
Invertebrates	98	0.01	49
Flowering plants	384	0.2	36

Source: Primack (2002).

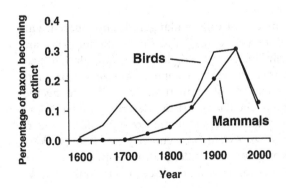

Fig. 1.1 Worldwide changes in extinction rates over time in mammals and birds (after Smith et al. 1995). *Extinction rates have generally increased for successive 50-year periods.*

without being recorded. Habitat loss will have resulted in extinctions of many undescribed species, especially of invertebrates, plants and microbes. Very few new species are likely to have evolved to replace those lost in this time.

The majority of recorded extinctions, and a substantial proportion of currently threatened species, are on islands (Table 1.1). For example, 81% of all extinct birds lived on islands, four-fold greater than the proportion of bird species that have lived on islands.

Extent of endangerment

18% of vertebrate animal species, 29% of invertebrates and 49% of plant species are classified as threatened

IUCN, the World Conservation Union, defines as **threatened** species with a high risk of extinction within a short time frame. These threatened species fall into the categories of critically endangered, endangered and vulnerable. In fish, amphibians, reptiles, birds and mammals IUCN classified 30%, 21%, 25%, 12% and 24% of assessed species as threatened. Of the 4763 species of mammals, 3.8% are critically endangered, 7.1% endangered and 13.0% vulnerable, while the remaining 76% are considered to be at lower risk. The situation is similar in invertebrates with 29% of assessed species classified as threatened./>

The situation in plants is if anything more alarming. IUCN classified 49% of plants as threatened, with 53% of mosses, 23% of gymnosperms, 54% of dicotyledons and 26% of monocotyledons threatened. There are considerable uncertainties about the data for all except mammals, birds and gymnosperms, as many species have not been assessed in the other groups. Estimates for microbes are not available, as the number of species in this groups is not known.

Projected extinction rates

Projections indicate greatly elevated extinction rates in the near future

With the continuing increase in the human population, and the anticipated impact on wildlife, there is a consensus that extinction rates are destined to accelerate markedly, typically by 1000-fold or more above the 'normal' background rates deduced from the fossilt record.

What is a threatened species?

Threatened species are those with a high risk of immediate extinction

The IUCN classifications of **critically endangered**, **endangered**, **vulnerable** and **lower risk** reflect degrees of risk of extinction. They are defined largely in terms of the rate of decline in population size, restriction in habitat area, the current population size and/or quantitatively predicted probability of extinction. Critically endangered species exhibit any one of the characteristics described under A–E in Table 1.2, i.e. $\geq$80% population size reduction over the last 10 years or three generations, or an extent of occupancy $\leq$100 square kilometres, or a stable population size $\leq$250 mature adults, or a probability of extinction $\geq$50% over 10 years or three generations, or some

Table 1.2 Designations of species into the critically endangered, endangered or vulnerable IUCN categories (IUCN 2002). A species conforming to any of the criteria A–E in the 'Critically endangered' column is defined as within that category. Similar rules apply to 'endangered' and 'vulnerable'

Criteria (any one of A–E)	Critically endangered	Endangered	Vulnerable
A. Actual or projected reduction in population size	80% decline over the last 10 years or three generations	50%	20%
B. Extent of occurrence or area of occupancy of	<100 km^2 <10 km^2 and any two of	<5000 km^2 <500 km^2	$<20\,000$ km^2 <2000 km^2
	(1) severely fragmented or known to exist at a single location, (2) continuing declines, and (3) extreme fluctuations	≤ 5 locations	≤ 10 locations
C. Population numbering and an estimated continuing decline, or population severely fragmented	<250 mature individuals	<2500	$<10\,000$
D. Population estimated to number	<50 mature individuals	<250	<1000
E. Quantitative analysis showing the probability of extinction in the wild	at least 50% within 10 years or three generations, whichever is the longer	20% in 20 years, or five generations	10% in 100 years

combination of these. For example, the critically endangered Javan rhinoceroses survive as only about 65 individuals in Southeast Asia and numbers continue to decline.

There are similar, but less threatening characteristics required to categorize species as endangered, or vulnerable. Species that do not conform to any of the criteria in Table 1.2 are designated as being at lower risk.

While there are many other systems used to categorize endangerment in particular countries and states, the IUCN provides the only international system and is the basis of listing species in the IUCN *Red Books* of threatened species. In general, we use the IUCN criteria throughout this book.

Importance of listing

Endangerment is the basis for legal protection of species. For example, most countries have Endangered Species Acts that provide legal protection for threatened species and usually require the formulation of recovery plans. In addition, threatened species are protected from trade by countries that have signed the Convention on International Trade in Endangered Species (CITES).

Listing of a species or sub-species as endangered provides a scientific foundation for national and international legal protection and may lead to remedial actions for recovery

What causes extinctions?

Human-associated factors

The primary factors contributing to extinction are directly or indirectly related to human impacts. The human population has grown exponentially and reached 6 billion on 12 October 1999. By 2050, the population is projected to rise to 8.9 billion, peaking at 10–11 billion around 2070 and then declining. This represents around a 75% increase above the current population. Consequently, human impacts on wild animals and plants will worsen in the near future.

Stochastic factors

Human-related factors often reduce populations to sizes where species are susceptible to accidental, or **stochastic**, effects. These are naturally occurring fluctuations experienced by small populations. They may have environmental, catastrophic, demographic, or genetic origins. Stochastic factors are discussed extensively throughout the book. Even if the original cause of population decline is removed, problems arising from small population size will persist unless these numbers recover.

Environmental stochasticity is random unpredictable variation in environmental factors, such as rainfall and food supply. **Demographic stochasticity** is random variation in birth and death rates and sex-ratios due to chance alone. **Catastrophes** are extreme environmental events due to tornadoes, floods, harsh winters, etc.

Genetic stochasticity encompasses the deleterious impacts of inbreeding, loss of genetic diversity and mutational accumulation on species. **Inbreeding** (the production of offspring from related parents), on average reduces birth rates and increases death rates (**inbreeding depression**) in the inbred offspring. Loss of **genetic diversity** reduces the ability of populations to adapt to changing environments via natural selection.

Environmental and demographic stochasticity and the impact of catastrophes interact with inbreeding and genetic diversity in their adverse effects on populations. If populations become small for any reason, they become more inbred, further reducing population size and increasing inbreeding. At the same time, smaller populations lose genetic variation (diversity) and consequently experience reductions in their ability to adapt and evolve to changing environments. This feedback between reduced population size, loss of genetic diversity and inbreeding is referred to as the **extinction vortex**. The complicated interactions between genetic, demographic and environmental factors can make it extremely difficult to identify the immediate cause(s) for any particular extinction event.

> The primary factors contributing to current extinctions are habitat loss, introduced species, over-exploitation and pollution. These factors are generated by humans, and related to human population growth

> Additional accidental (stochastic) environmental, catastrophic, demographic and genetic factors increase the risk of extinction in small populations

What is conservation genetics?

Conservation genetics is the use of genetic theory and techniques to reduce the risk of extinction in threatened species. Its longer-term goal is to preserve species as dynamic entities capable of coping with environmental change. Conservation genetics is derived from evolutionary genetics and from the quantitative genetic theory that underlies selective breeding of domesticated plants and animals. However, these theories generally concentrate on large populations where the genetic constitution of the population is governed by deterministic factors (selection coefficients, etc.). Conservation genetics is now a discrete discipline focusing on the consequences arising from reduction of once-large, outbreeding, populations to small units where stochastic factors and the effects of inbreeding are paramount.

The field of conservation genetics also includes the use of molecular genetic analyses to elucidate aspects of species' biology relevant to conservation management.

Major issues include:

- the deleterious effects of inbreeding on reproduction and survival (**inbreeding depression**)
- loss of **genetic diversity** and ability to evolve in response to environmental change (loss of **evolutionary potential**)
- fragmentation of populations and reduction in gene flow
- random processes (**genetic drift**) overriding natural selection as the main evolutionary process
- accumulation and loss (**purging**) of deleterious mutations
- genetic management of small captive populations and the adverse effect of adaptation to the captive environment on reintroduction success
- resolution of taxonomic uncertainties
- definition of management units within species
- use of molecular genetic analyses in **forensics** and elucidation of aspects of species biology important to conservation.

Some examples are given below.

Reducing extinction risk by minimizing inbreeding and loss of genetic diversity

Many small, threatened populations are inbred and have reduced levels of genetic diversity. For example, the endangered Florida panther suffers from genetic problems as evidenced by low genetic diversity, and inbreeding-related defects (poor sperm and physical abnormalities). To alleviate these effects, individuals from its most closely related sub-species in Texas have been introduced into this population. Captive populations of many endangered species (e.g. golden lion tamarin) are managed to minimize loss of genetic diversity and inbreeding.

Florida panther

Identifying species or populations at risk due to reduced genetic diversity

Asiatic lions exist in the wild only in a small population in the Gir Forest in India and have very low levels of genetic diversity. Consequently, they have a severely compromised ability to evolve, as well as being susceptible to demographic and environmental risks. The

Wollemi pine

Red-cockaded woodpecker

Velvet worm

recently discovered Wollemi pine, an Australian relict species previously known only from fossils, contains no genetic diversity among individuals at several hundred loci. Its extinction risk is extreme. It is susceptible to a common die-back fungus and all individuals that were tested were similarly susceptible. Consequently, a program has been instituted that involves keeping the site secret, quarantine, and the propagation of plants in other locations.

Resolving fragmented population structures

Information regarding the extent of gene flow among populations is critical to determining whether a species requires human-assisted exchange of individuals to prevent inbreeding and loss of genetic diversity. Wild populations of the red-cockaded woodpecker are fragmented, causing genetic differentiation among populations and reduction of genetic diversity in small populations. Consequently, part of the management of this species involves moving (translocating) individuals into small populations to minimize the risks of inbreeding and loss of genetic diversity.

Resolving taxonomic uncertainties

The taxonomic status of many invertebrates and lower plants is frequently unknown. Thus, an apparently widespread and low-risk species may, in reality, comprise a complex of distinct species, some rare or endangered. In Australia, tarantula spiders are apparently widespread in northern tropical forests and are collected for trade. However, experts can identify even pet-shop specimens as undescribed species, some of which may be native only to restricted regions. They may be driven to extinction before being recognized as threatened species. Similar studies have shown that Australia is home to well over 100 locally distributed species of velvet worms (*Peripatus*) rather than the seven widespread morphological species previously recognized. Even the unique New Zealand tuatara reptile has been shown to consist of two, rather than one species.

Equally, genetic markers may reveal that populations thought to be threatened actually belong to common species, and are attracting undeserved protection and resources. Molecular genetic analyses have shown that the endangered colonial pocket gopher from Georgia, USA is indistinguishable from the common pocket gopher in that region.

Defining management units within species

Populations within species may be adapted to somewhat different environments and be sufficiently differentiated to deserve management as separate units. Their hybrids may be at a disadvantage, sometimes even displaying partial reproductive isolation. For example, coho salmon (and many other fish species) display genetic differentiation among populations from different geographic locations. These show evidence of adaptation to different conditions (morphology, swimming ability and age at maturation). Thus, they should be managed as separate populations.

Coho salmon

Detecting hybridization

Many rare species of plants, salmonid fish and canids are threatened with being 'hybridized out of existence' by crossing with common species. Molecular genetic analyses have shown that the critically endangered Ethiopian wolf (simian jackal) is subject to hybridization with local domestic dogs.

Non-intrusive sampling for genetic analyses

Many species are difficult to capture, or are badly stressed in the process. DNA can be obtained from hair, feathers, sloughed skin, faeces, etc. in non-intrusive sampling, the DNA amplified and genetic studies completed without disturbing the animals. For example, the critically endangered northern hairy-nosed wombat is a nocturnal burrowing marsupial which can only be captured with difficulty. They are stressed by trapping and become trap-shy. Sampling has been achieved by placing adhesive tape across the entrances to their burrows to collect hair when the animals exit their burrows. DNA from non-invasive sampling can be used to identify individuals, determine mating patterns and population structure, and measure levels of genetic diversity.

Northern hairy-nosed wombat

Defining sites for reintroduction

Molecular analyses may provide additional information on the historical distribution of species, expanding possibilities for conservation action. For ecological reasons, reintroductions should preferably occur within a species' historical range. The northern hairy-nosed wombat exists in a single population of approximately 100 animals at Clermont in Queensland, Australia. DNA samples obtained from museum skins identified an extinct wombat population at Deniliquin in New South Wales as belonging to this species. Thus, Deniliquin is a potential site for reintroduction. Similarly, information from genotyping DNA from sub-fossil bones has revealed that the endangered Laysan duck previously existed on islands other than its present distribution in the Hawaiian Islands.

Choosing the best populations for reintroduction

Island populations are considered to be a valuable genetic resource for re-establishing mainland populations, particularly in Australia and New Zealand. However, molecular genetic analyses revealed that the black-footed rock wallaby population on Barrow Island, Australia (a potential source of individuals for reintroductions onto the mainland) has extremely low genetic variation and reduced reproductive rate (due to inbreeding). Some numerically smaller and more endangered mainland populations are genetically healthier and are therefore a more suitable source of animals for reintroductions to other mainland localities. Alternatively, the pooling of several different

Black-footed rock wallaby

island populations of this wallaby could provide a genetically healthy population suitable for reintroduction purposes.

Forensics

Molecular genetic methods are widely applied to provide forensic evidence for litigation. These include the detection of illegal hunting and collection. Sale for consumption of meat from threatened whales has been detected by analysing samples in Japan and South Korea. Mitochondrial DNA sequences showed that about 9% of the whale meat on sale came from protected species, rather than from the minke whales that are taken legally. Methods have been devised to identify species of origin using small amounts of DNA from shark fins and work is in progress to identify tiger bones in Asian medicines.

Humpback whale

Understanding species biology

Many aspects of species biology can be determined using molecular genetic analyses. For example, mating patterns and reproduction systems are often difficult to determine in threatened species. Studies using genetic markers established that loggerhead turtle females mate with several males. Mating systems in many plants have been established using genetic analyses. Birds are often difficult to sex, resulting in several cases where two birds of the same sex were placed together to breed. Molecular genetic methods are now available to sex birds without resorting to surgery. Paternity can be determined in many species, including chimpanzees. Endangered Pyrenean brown bears are nocturnal, secretive and dangerous. Methods have been devised to census these animals, based upon DNA from hair and faeces. Individuals can be sexed and uniquely identified.

Dispersal and migration patterns are often critical to species survival prospects. These are difficult to determine directly, but can be inferred using genetic analyses.

Each of these aspects will be explained in later chapters.

SUGGESTED FURTHER READING

Frankham, R., J. D. Ballou & D. A. Briscoe. 2002. *Introduction to Conservation Genetics*. Cambridge University Press, Cambridge, UK. Comprehensive textbook of conservation genetics. Chapter 1 has an extended treatment of these topics, plus references.

IUCN. 2002. *Red List of* Threatened Species. Website with full details of the international recognized IUCN categorization system for designating threatened species, plus up-to-date summaries of the proportions of species threatened in different major groups and categorization lists for animals and links to a plant database.

Leakey, R. & R. Lewin. 1995. *The Sixth Extinction: Biodiversity and its Survival*. Phoenix, London. Account of the biodiversity crisis written for a general audience.

Meffe, G. K. & C. R. Carroll. 1997. *Principles of Conservation Biology*, 2nd edn. Sinauer, Sunderland, MA. Basic textbook in conservation biology, with a reasonable coverage of genetic issues.

Primack, R. B. 2002. *Essentials of Conservation Biology*, 3rd edn. Sinauer, Sunderland, MA. Basic textbook in conservation biology with a good, but limited coverage of genetic issues.

Chapter 2

Genetic diversity

Genetic diversity is required for populations to adapt to environmental change. Large populations of naturally outbreeding species usually have extensive genetic diversity, but it is typically reduced in endangered species

Terms

Allelic diversity, allozyme, amplified fragment length polymorphism (AFLP), DNA fingerprint, electrophoresis, fitness, genome, Hardy–Weinberg equilibrium, heritability, hermaphrodite, heterozygosity, intron, locus, microsatellite, mitochondrial DNA (mtDNA), monomorphic, mutation load, outbreeding, polymerase chain reaction (PCR), polymorphism, probe, quantitative character, quantitative genetic variation, quantitative trait loci (QTL), random amplified polymorphic DNA (RAPD), restriction fragment length polymorphism (RFLP), silent substitution, single nucleotide polymorphism (SNP), synonymous substitutions

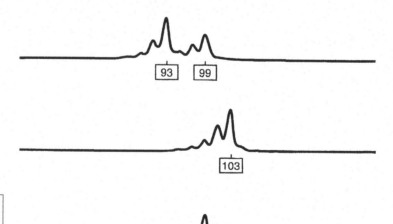

A Galápagos tortoise and output from a DNA sequencing machine illustrating genetic diversity at a microsatellite locus among individuals in this species

Importance of genetic diversity

IUCN, the premier international conservation body, recognizes the need to conserve genetic diversity as one of three global conservation priorities. There are two major and interrelated issues. First, environmental change is a continuous process and genetic diversity is required for populations to evolve to adapt to such change. Second, loss of genetic diversity is usually associated with inbreeding and overall reduction in reproduction and survival (**fitness**).

Captive breeding and wildlife management programs typically recognize the importance of minimizing loss of genetic diversity and inbreeding. Management action for captive populations includes consulting pedigrees when establishing matings or choosing individuals to reintroduce into the wild. Levels of genetic diversity are analysed and monitored in wild populations of endangered species, and gene flow between isolated wild populations may be augmented.

This chapter addresses the basis of the two major issues concerning genetic diversity, defines what it is, describes methods for measuring it, and reviews evidence on its extent in non-endangered and endangered species.

> Genetic diversity is the raw material upon which natural selection acts to bring about adaptation and evolution to cope with environmental change. Loss of genetic diversity reduces evolutionary potential and is also associated with reduced reproductive fitness

What is genetic diversity?

Genetic diversity is manifested by differences in many characters, including eye, skin and hair colour in humans, colour and banding patterns of snail shells, flower colours in plants, and in the proteins, enzymes and DNA sequences of almost all organisms.

Genes are sequences of nucleotides in a particular region (locus) of a DNA molecule. Genetic diversity represents slightly different sequences. In turn, DNA sequence variants may be expressed in amino acid sequence differences in the protein the locus codes for. Such protein variation may result in functional biochemical, morphological or behavioural dissimilarities that cause differences in reproductive rate, survival or behaviour of individuals.

The terminology used to describe genetic diversity is defined in Table 2.1. Genetic diversity is typically described using **polymorphism**, average **heterozygosity**, and **allelic diversity**. For example, in African lions 6 of 26 protein-coding loci (23%) were variable (polymorphic), 7.1% of loci were heterozygous in an average individual, and there was an average of 1.27 alleles per locus (allelic diversity), as assessed by allozyme electrophoresis (see below). These levels of genetic diversity are typical of electrophoretic variation for non-threatened mammals. By contrast, endangered Asiatic lions have low genetic diversity.

> Genetic diversity is the variety of alleles and genotypes present in the group under study (population, species or group of species)

Asiatic lion

> The genetic composition of a population is usually described in terms of allele frequencies, number of alleles and heterozygosity

Measuring genetic diversity

Molecular techniques such as allozyme electrophoresis or microsatellite typing (see below) measure genetic diversity at individual loci.

Table 2.1 Terminology used to describe genetic diversity

Locus (plural **loci**) The site on a chromosome at which a particular gene is located. The nucleotide sequence at a locus may code for a particular structure or function, e.g. the segment of DNA coding for the alcohol dehydrogenase enzyme is a separate locus from those coding for haemoglobins. Molecular loci, such as microsatellites (see below), are simply segments of DNA that may have no functional products.

Alleles Different variants of the nucleotide sequence at the same locus (gene) on homologous chromosomes, e.g. A_1, A_2, A_3, A_4, etc.

Genotype The combination of alleles present at a locus in an individual, e.g. A_1A_1, A_1A_2 or A_2A_2. Genotypes are heterozygous (A_1A_2) or homozygous (A_1A_1 or A_2A_2).

Genome The complete genetic material of a species, or individual; the entire DNA nucleotide sequence, including all of the loci and all of the chromosomes.

Homozygote An individual with two copies of the same allele at a locus, e.g. A_1A_1.

Heterozygote An individual with two different alleles at a locus, e.g. A_1A_2.

Allele frequency The relative frequency of a particular allele in a population (often referred to as gene frequency). For example, if a population of a diploid species has 8 A_1A_1 individuals and 2 A_1A_2 individuals, then there are 18 copies of the A_1 allele and 2 of the A_2 allele. Thus, the A_1 allele has a frequency of 0.9 and the A_2 allele a frequency of 0.1.

Polymorphic The presence in a species of two or more alleles at a locus, e.g. A_1 and A_2. Polymorphic loci are usually defined as having the most frequent allele at a frequency of less than 0.99, or less than 0.95 (to minimize problems with different sample sizes).

Monomorphic A locus in a population is monomorphic if it has only one allele present, e.g. A_1. All individuals are homozygous for the same allele. Lacking genetic diversity.

Proportion of loci polymorphic (P) Number of polymorphic loci / total number of loci sampled. For example, if 3 of 10 sampled loci are polymorphic, and 7 are monomorphic,

$$P = \frac{3}{10} = 0.3$$

Average heterozygosity (H) Sum of the proportions of heterozygotes at all loci / total number of loci sampled. For example, if the proportions of individuals heterozygous at 10 loci in a population are 0.2, 0.4, 0.1, 0, 0, 0, 0, 0, 0, and 0, then

$$H = \frac{(0.2 + 0.4 + 0.1 + 0 + 0 + 0 + 0 + 0 + 0 + 0)}{10} = 0.07$$

Typically, expected heterozygosities (see below) are reported, as they are less sensitive to sample size than observed heterozygosities. In random mating populations, observed and expected heterozygosities are usually similar.

Allelic diversity (A) average number of alleles per locus
For example, if the number of alleles at 10 loci are 2, 3, 2, 1, 1, 1, 1, 1, 1 and 1, then

$$A = \frac{(2 + 3 + 2 + 1 + 1 + 1 + 1 + 1 + 1 + 1)}{10} = 1.4$$

Table 2.2 | Numbers and frequencies for each of the genotypes at an egg-white protein locus in eider ducks from Scotland. Individuals were genotyped by protein electrophoresis. F refers to the faster-migrating allele and S to the slower

| | Genotypes | | | |
	FF	FS	SS	Total
Numbers	37	24	6	67
Genotype frequencies	0.552	0.358	0.090	1.00

Source: Milne & Robertson (1965).

Eider ducks

The information we collect provides the numbers of each genotype at a locus. This is illustrated for an egg-white protein locus in Scottish eider ducks, a species that was severely depleted due to harvest of feathers for bedding (Table 2.2). Genotype frequencies are simply calculated from the proportion of the total sample of that type (e.g. genotype frequency of FF = 37/67 = 0.552).

The information is usually reported in the form of allele frequencies, rather than genotype frequencies. We use the letters p and q to represent the relative frequencies for the two alleles at the locus. The frequency of the F allele (p) is simply the proportion of all alleles examined which are F. Note that we double the number of each homozygote, and the total, as the ducks are diploid (each bird has inherited one copy of the locus from each of its parents).

$$p = \frac{(2 \times FF) + FS}{2 \times Total} \tag{2.1}$$

The calculation in Example 2.1 shows that 73% of the alleles at this locus are the F allele and 27% are S.

Example 2.1 | Calculation of F and S allele frequencies at an egg-white protein locus in eider ducks

The frequency for the F allele (p) is obtained as follows:

$$p = \frac{(2 \times 37) + 24}{(2 \times 67)} = 0.73$$

and that for S (q) as

$$q = \frac{(2 \times 6) + (1 \times 24)}{(2 \times 67)} = 0.27$$

The sum of their frequencies is 1.00, i.e.

$$p + q = 0.73 + 0.27 = 1.00.$$

Allele frequencies may also be reported as percentages.

The extent of genetic diversity at a locus is expressed as heterozygosity. **Observed heterozygosity** (H_o) is simply the proportion of the sampled individuals that are heterozygotes. For example, the observed

Heterozygosity is the measure most commonly used to characterize genetic diversity for single loci frequency of heterozygotes at the egg-white protein locus in eider ducks is $24/67 = 0.36$ (Table 2.2). When we are comparing the extent of genetic diversity among populations or species we typically use average heterozygosity (Table 2.1).

Allelic diversity

Allelic diversity is also used to characterize genetic diversity

The average number of alleles per locus (allelic diversity) is also used to characterize the extent of genetic diversity. For example, there are two alleles at the locus determining egg-white protein differences in eider ducks and three alleles at the microsatellite locus in Laysan finches, described in Example 2.2 below. When more than one locus is studied, allelic diversity (A) is the number of alleles averaged across loci (Table 2.1). For example, African lions have a total of 33 alleles over the 26 allozyme loci surveyed, so $A = 33/26 = 1.27$.

We now examine the factors that influence the frequencies of alleles and genotypes in a population, and the relationship between allele and genotype frequencies under the assumption of random union of gametes (random mating is equivalent to this).

Hardy–Weinberg equilibrium

In large, random mating populations, allele and genotype frequencies at an autosomal locus attain equilibrium after one generation when there are no perturbing forces (no mutation, migration or selection)

Let us begin with the simplest case – that of a large population where mating is random and there is no mutation, migration or selection. In this case, allele and genotype frequencies attain an equilibrium after just one generation. The equilibrium is named the **Hardy–Weinberg equilibrium**, after its discoverers. While the Hardy–Weinberg equilibrium is very simple, it is crucial in conservation and evolutionary genetics. It provides a basis for detecting deviations from random mating, testing for selection, modelling the effects of inbreeding and selection, and estimating the allele frequencies at loci showing dominance.

This simple case can be presented as a mathematical model that shows the relationship between allele and genotype frequencies. Assume that we are dealing with a locus with two alleles A_1 and A_2 at relative frequencies of p and q ($p + q = 1$) in a large random mating population. Imagine **hermaphroditic** (both sperm and eggs released by each individual) marine organisms shedding their gametes into the water, where sperm and eggs unite by chance (Table 2.3). Since the allele frequency of A_1 in the population is p, the frequency of sperm or eggs carrying that allele is also p. The probability of a sperm carrying A_1 uniting with an egg bearing the same allele, to produce an A_1A_1 zygote, is therefore $p \times p = p^2$ and the probability of an A_2 sperm fertilizing an A_2 egg, to produce a A_2A_2 zygote is, likewise, $q \times q = q^2$. Heterozygous zygotes can be produced in two ways, it does not matter which gamete contributes which allele, and their expected frequency is therefore $2 \times p \times q = 2pq$. Consequently, the expected genotype frequencies for A_1A_1, A_1A_2 and A_2A_2 zygotes are p^2, $2pq$ and q^2, respectively. These are the **Hardy–Weinberg equilibrium** genotype frequencies.

Table 2.3 Genotype frequencies resulting from random union of gametes at an autosomal locus

		Ova	
		A_1	A_2
		p	q
Sperm	A_1 p	p^2 A_1A_1	pq A_1A_2
	A_2 q	pq A_2A_1	q^2 A_2A_2

The resulting genotype frequencies in progeny are

A_1A_1	A_1A_2	A_2A_2
p^2	$2pq$	q^2

These are the Hardy–Weinberg equilibrium genotype frequencies. Note that the genotype frequencies sum to unity,

i.e. $p^2 + 2pq + q^2 = (p + q)^2 = 1$.

If the frequencies of the alleles A_1 and A_2 are 0.9 and 0.1, then the Hardy–Weinberg equilibrium genotype frequencies are:

A_1A_1	A_1A_2	A_2A_2	Total
0.9^2	$2 \times 0.9 \times 0.1$	0.1^2	1.0
0.81	0.18	0.01	1.0

The frequencies of the two alleles have not changed, indicating that allele frequencies are in **equilibrium**. Consequently, allele and genotype frequencies are at equilibrium after one generation of random mating and remain so in perpetuity in the absence of other influences.

Hardy–Weinberg equilibrium is expected for all loci, except for those located on sex chromosomes. These sex-linked loci have different doses of loci in males and females and have Hardy–Weinberg equilibria that differ from those for non sex-linked (autosomal loci) loci.

The relationship between allele and genotype frequencies according to the Hardy–Weinberg equilibrium is shown in Fig. 2.1. This illustrates two points. First, the frequency of heterozygotes cannot be greater than 0.5 (50%) for a locus with two alleles. This occurs when both the alleles have frequencies of 0.5. Second, when an allele is rare, most of its alleles are in heterozygotes, while most are in homozygotes when it is at a high frequency.

To obtain the Hardy–Weinberg equilibrium we assumed:

- a large population size
- a closed population (no migration)
- no mutation
- normal Mendelian segregation of alleles
- equal fertility of parent genotypes
- random union of gametes

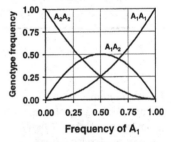

Fig. 2.1 Relationship between genotype frequencies and allele frequencies in a population in Hardy–Weinberg equilibrium.

Genotype frequencies for most loci usually agree with Hardy–Weinberg genotype frequency expectations in large naturally outbreeding populations

- equal fertilizing capacity of gametes
- equal survival of all genotypes

The genotype frequencies for the eider duck egg-white protein locus are compared with the Hardy–Weinberg equilibrium frequencies in Table 2.4. Values of p and q, calculated previously, are used to calculate p^2, $2pq$ and q^2. These frequencies are then multiplied by the total number (67) to obtain expected numbers for the three genotypes.

The observed numbers for each genotype are very close to the numbers expected from the Hardy–Weinberg equilibrium. In general, agreement with expectations is found for most loci in large naturally **outbreeding** populations (more or less random mating). This does not mean that the loci are not subject to mutation, migration, selection and sampling effects, only that these effects are often too small to be detectable with realistic sample sizes.

Table 2.4 Comparison of observed genotype frequencies with Hardy–Weinberg equilibrium expectations for the eider duck egg-white protein locus

| | Genotypes | | | |
	FF	FS	SS	Total
Observed numbers (O)	37	24	6	67
Expected frequencies	p^2	$2pq$	q^2	1.0
	0.73^2	$2 \times 0.73 \times 0.27$	0.27^2	1.0
	0.5329	0.3942	0.0729	1.0
Expected numbers (E) (expected frequency $\times$ 67)	35.7	26.4	4.9	67

Deviations from Hardy–Weinberg equilibrium

When any of the assumptions underlying the Hardy–Weinberg equilibrium (no migration, selection or mutation, and random union of gametes) are violated, then deviations from the equilibrium genotype frequencies will occur. Thus, the Hardy–Weinberg equilibrium provides a null hypothesis that allows us to detect if the population has non-random mating, migration, or selection. We deal with these in this and later chapters.

Expected heterozygosity

Genetic diversity at a single locus is characterized by expected heterozygosity, observed heterozygosity and allelic diversity. For a locus with two alleles at frequencies of p and q, the **expected heterozygosity** is $H_e = 2pq$ (also called **gene diversity**). When there are more than two alleles, it is simpler to calculate expected heterozygosity as one minus the sum of the squared allele frequencies:

$$H_e = 1 - \sum_{i=1}^{\#\,\text{alleles}} p_i^2 \qquad (2.2)$$

where p_i is the frequency of the ith allele. H_e is usually reported in preference to observed heterozygosity.

Example 2.2 illustrates the calculation of expected heterozygosity for a microsatellite locus in Laysan finches. The Hardy–Weinberg expected heterozygosity is 0.663, based upon the allele frequencies at this locus.

Example 2.2 | Calculating expected heterozygosity for a microsatellite locus in the endangered Laysan finch (Tarr et al. 1998)

The allele frequencies for three alleles are 0.364, 0.352 and 0.284, respectively. Consequently, the Hardy–Weinberg expected heterozygosity is

$$H_e = 1 - (0.364^2 + 0.352^2 + 0.284^2) = 1 - (0.1325 + 0.1239 + 0.0807)$$

$$= 0.663$$

The observed and expected heterozygosities of 0.659 and 0.663 at this locus are very similar.

To assess the evolutionary potential of a species, it is necessary to estimate the extent of genetic diversity across all loci in the genome. Information on a single locus is unlikely to accurately depict genetic diversity for all loci in a species. For example, mammals have around 35 000 functional loci. Consequently, genetic diversity measures (H_o, H_e) are averaged over a random sample of many loci. For example, the average heterozygosity detected using protein electrophoresis for 26 loci in African lions is 7.1%, as described above.

> Average heterozgyosity over several loci is used to characterize genetic diversity in a species

Estimating the frequency of a recessive allele

It is not possible to determine the frequency of an allele at a locus showing dominance using the allele counting method outlined above, as dominant homozygotes (AA) cannot be distinguished phenotypically from heterozygotes (Aa). However, the Hardy–Weinberg equilibrium provides a means for estimating the frequencies of such alleles. Recessive homozygotes (aa) are phenotypically distinguishable and have an expected frequency of q^2 for a locus in Hardy–Weinberg equilibrium. Thus, the recessive allele frequency can be estimated as the square root of this frequency. For example, the frequency of chondrodystrophic dwarfism in the endangered California condor is 0.03, so the recessive allele causing the condition has an estimated frequency of $\sqrt{0.03} = 0.17$. This is a surprisingly high frequency for a recessive lethal allele, but is not uncommon in other populations derived from very few founders, including populations of other endangered species.

> The Hardy–Weinberg equilibrium provides a means for estimating the frequencies of recessive alleles in random mating populations

Since there are several assumptions underlying this method of estimating q (random mating, no selection or migration), it should never be used for loci where all genotypes can be distinguished.

Extent of genetic diversity

Large populations of naturally outbreeding species usually have extensive genetic diversity

Typically, large populations of outbreeding species contain vast amounts of genetic diversity. This is manifested in morphological, behavioural and physiological variations in most populations. This variation is composed of both non-genetic-based variation, due to environmental influences on individuals, and genetic-based variation due to differences in alleles and heterozygosity at many loci. An example of the large amount of genetic diversity inherent in a species is the variety of dog breeds that have been produced from the wolf genome (Fig. 2.2). With the exception of a few mutations, the variety of dog breeds reflects the extent of genetic diversity that was present in the ancestral wolves.

Fig. 2.2 Diversity of dog breeds. All derive from the gray wolf.

gray wolf

Genetic diversity can be measured at a number of different levels. This includes diversity in measurable characters (quantitative variation), the visible direct effects of deleterious alleles, variation in proteins, and direct measurement of variation in DNA sequences.

Quantitative variation

The characters of most importance in conservation are quantitative characters

The characters of most importance in conservation are those that determine the health and reproductive fitness of organisms. These include such attributes as number of offspring produced, seed yield in plants, mating ability, longevity, growth rate, predator avoidance behaviours, body weight, heights, strength, etc. Collectively, these kinds of traits are called **quantitative characters**. Virtually all quantitative characters in outbreeding species exhibit genetic diversity. For example, genetic diversity has been found for reproductive characters (egg production in chickens, number of offspring in sheep, mice, pigs and fruit flies, and seed yield in plants, etc.), for growth rate in size (in cattle, pigs, mice, chickens, fruit flies and plants), for chemical composition (fat content in animals, protein and oil levels in maize), for behaviour (in insects and mammals) and for disease resistance in plants and animals. Quantitative characters are determined by many loci (**quantitative trait loci**, or QTL).

Genetic diversity for a quantitative character is typically determined by measuring similarities in the trait among many related individuals and determining the proportion of the phenotypic variation that is heritable (**heritability**). We will discuss this in Chapter 3.

Deleterious alleles

The extent of diversity in populations attributable to deleterious alleles is critical in conservation biology because these alleles reduce viability and reproductive fitness when they become homozygous through inbreeding. Deleterious alleles are constantly generated by mutation and removed by selection. Consequently, all outbred populations contain deleterious rare alleles (**mutation load**). Typically, these occur at frequencies of less than 1%. Rare human genetic syndromes, such as phenylketonuria, albinism and Huntington disease are examples. Equivalent syndromes exist in wild populations of plants and animals. For example, mutations leading to a lack of chlorophyll are found in many species of plants. A range of genetically based defects has been described in many endangered animals (dwarfism in California condors, vitamin E malabsorption in Przewalski's horse, undescended testes and fatal heart defects in Florida panthers, lack of testis in koalas and hairlessness in red-ruffed lemurs).

> All outbred populations contain a 'load' of rare deleterious alleles that can be exposed by inbreeding

Proteins

The first measures of genetic diversity at the molecular level were made in 1966 by studying allelic variation at loci coding for soluble proteins. The technique used to distinguish variants is **electrophoresis**, which separates molecules according to their net charge and molecular mass in an electrical potential gradient (Box 2.1). However, only about 30% of changes in DNA result in charge changes in the proteins, so this technique significantly underestimates the full extent of genetic diversity.

Protein electrophoresis is typically conducted using samples of blood, liver or kidney in animals, or leaves and root tips in plants, as these contain ample amounts and varieties of soluble proteins. Consequently, animals must be captured to obtain blood samples, or killed to obtain liver or kidney samples. These are undesirable practices for endangered species. Soluble proteins are relatively fragile molecules and protein techniques, unlike DNA techniques, require fresh or fresh-frozen samples.

> Extensive information on genetic diversity has been obtained using electrophoretic separation of proteins

Box 2.1 | Measuring genetic diversity in proteins using allozyme electrophoresis

The sequence of amino acids making up a protein is determined by the sequence of bases in the DNA coding for that protein. A proportion of base changes result in amino acid change at the corresponding position in the protein. As five of the 20 naturally occurring amino acids are electrically charged (lysine, arginine and histidine [+], glutamic acid and aspartic acid [−]), about 30% of the base substitutions result in changes to the electrical charge of the resultant protein. These changes can be detected by separating the proteins in an electrical potential gradient and subsequently visualising them using a locus-specific histochemical stain. This process is termed **allozyme** electrophoresis. For example, if the DNA in alleles at a locus have the sequences:

	DNA	DNA with base substitution
DNA strand	....TAC GAA CTG CAA....	....TAC GAA C**C**G CAA....
mRNA	....AUG CUU GAC GUU....	....AUG CUU G**G**C GUU....
amino acid sequence	...met– leu– asp– val....	....met– leu– **gly**– val....

the protein on the right will migrate more slowly towards the anode in an electrical potential gradient, as a consequence of the substitution of uncharged glycine (gly) amino acid for negatively charged aspartic acid (asp).

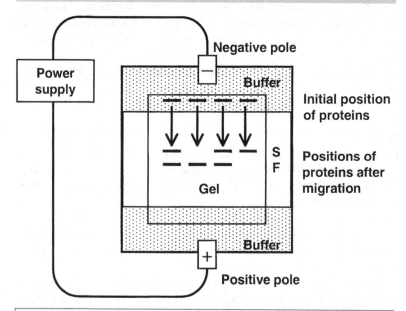

A gel electrophoresis apparatus (after Hedrick 1983). Soluble protein extracts are placed in spaced positions across the top of the gel. An electrical potential gradient applied to the gel causes the proteins to migrate through the gel. Proteins coded for by the same genetic locus but with different charges migrate to different positions (F–fast vs. S–slow), allowing identification of the different alleles at the locus. Proteins from specific loci are usually detected by their unique enzymatic activity, using a histochemical stain.

Table 2.5 | Allozyme genetic diversity in different taxa. H is the average heterozygosity within populations

	H
Vertebrates	
Total	0.064
Mammals	0.054
Birds	0.054
Reptiles	0.090
Amphibians	0.094
Fish	0.054
Invertebrates	
Total	0.113
Insects	0.122
Crustaceans	0.063
Molluscs	0.121
Plants	
Total	0.113
Gymnosperms	0.160
Monocotyledons	0.144
Dicotyledons	0.096

Sources: Hamrick & Godt (1989); Ward *et al.* (1992).

The first analyses of electrophoretic variation, in humans and fruit flies, revealed surprisingly high levels of genetic diversity. Similar results are found for most species with large population sizes. For example, in humans (based on 104 loci) 32% of loci are polymorphic with an average heterozygosity of 6%. Table 2.5 summarizes allozyme heterozygosities for several major taxonomic groups. Average heterozygosity within species (H) is lower in vertebrates (6.4%) than in invertebrates (11.3%) or plants (11.3%), possibly due to lower population sizes in vertebrates.

> There is extensive genetic diversity at protein-coding loci in most large populations of outbred species. On average 28% of loci are polymorphic and 7% of loci are heterozygous in an average individual, as assessed by electrophoresis

DNA

Collecting DNA samples for measuring genetic diversity

Any biological material containing DNA can be used to measure genetic diversity with modern molecular techniques. For example, shed hair, skin, feathers, faeces, urine, egg shell, fish scales, blood, tissues, saliva and semen are suitable. Museum skins and preserved tissues provide adequate material and even fossils may be genotyped. The only requirements are that the sample contains some undegraded DNA and that it is not contaminated with DNA from other individuals or closely related species.

DNA amplification using PCR

Many current methods of measuring DNA diversity rely on the **polymerase chain reaction (PCR)** which allows laboratory amplification of specific DNA sequences, often from small initial samples (Fig. 2.3).

> Genotyping of individuals can be done following non-invasive or 'remote' sampling and PCR amplification of DNA

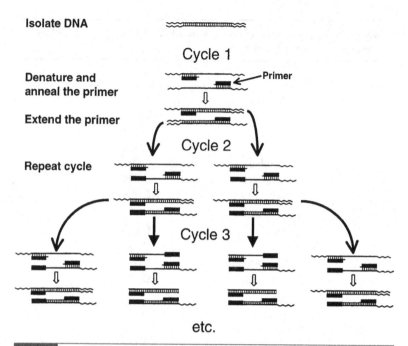

Isolate DNA

Cycle 1

Denature and
anneal the primer

Primer

Extend the primer

Cycle 2

Repeat cycle

Cycle 3

etc.

Fig. 2.3 Non-invasive sampling of DNA and use of the polymerase chain reaction (PCR) to amplify DNA. PCR is used to amplify (generate multiple copies of) DNA from tiny samples. PCR is essentially a test-tube version of natural DNA replication, except that it only replicates the DNA region of interest. DNA is extracted and purified from the biological sample and added to a reaction mix containing all the necessary reagents. These include DNA oligonucleotide primers, a heat-resistant DNA replicating enzyme (*Taq* polymerase), magnesium, the four DNA nucleotides and buffer. The primers are homologous to the conserved DNA sequences on either side of (flanking) the DNA sequence to be amplified (i.e. the locus of interest). The *Taq* polymerase enzyme replicates DNA, the nucleotides are the building blocks of the new DNA strands and magnesium and buffer are required for the enzyme to work.

Repeated temperature cycles are used to denature the DNA (separate the strands), allow the DNA primers to attach to the flanking sequences (anneal), and to replicate the DNA sequence between the two primers (extend). Each cycle doubles the quantity of DNA of interest.

A major advantage of measuring DNA variation, as opposed to protein variation, is that sampling can often be taken non-invasively, and genotypes identified following DNA amplification. Since extremely small samples of DNA (as little as the content of a single cell) can be amplified millions of times by PCR, only minute biological samples are now needed to conduct molecular genetic analyses. This contrasts with protein electrophoresis where animals must be caught or killed to obtain samples. Consequently, the development of 'remote' sampling methods has been a major advance for species of conservation concern.

To amplify a DNA segment of interest, specific invariant (conserved) sequences on either side of the segment of interest must be identified to design primers for the PCR reaction. The segment to be amplified is defined by, and lies between the primers. Copies of these sequences are synthesized (oligonucleotides) and used in the PCR reaction. Primer sequences can often be deduced from published sequence information for mitochondrial DNA (mtDNA), but must frequently be developed anew for nuclear loci, especially for microsatellites (see below). Primers developed for one species may also work in a closely related species. For example, human primers usually work in chimpanzees, and some of the primers from domestic ruminants work in the endangered Arabian oryx.

There is an array of techniques available for directly, or indirectly measuring DNA base sequence variation, and new methods are regularly being devised (Box 2.2). DNA sequencing is routinely conducted, especially for taxonomic purposes. Microsatellites (variable number short tandem repeats), have become the marker of choice for population studies. Microsatellites have advantages over other methods to measure DNA variation as they are highly variable, individual genotypes can be directly inferred, and individuals can be typed following non-invasive sampling. They have the disadvantage that the primers must be developed anew for each species, although primers from closely related species will often work in both species.

> A wide array of methods is available for measuring diversity in DNA base sequences, with microsatellites being the method currently favoured

---

| **Box 2.2** | **Techniques for measuring genetic diversity in DNA** |

MICROSATELLITES: SIMPLE SEQUENCE REPEATS (SSR) OR SHORT TANDEM REPEATS (STR)

Microsatellite loci are tandem repeats of short DNA motifs, typically 1–5 bases in length. For example, the DNA base sequence CA with 7 and 9 repeats are shown. CA repeats are found in many species. The number of microsatellite repeats is highly variable due to 'slippage' during DNA replication. The double-stranded DNA sequence of three genotypes, two different homozygotes and a heterozygote are illustrated below, along with their banding patterns following electrophoresis on a sequencing gel. **X** and **Y** are invariant (conserved) DNA sequences (primer sites) flanking the microsatellite repeat.

A_1A_1	A_1A_2	A_2A_2
XCACACACACACACA**Y**	**X**CACACACACACACACACA**Y**	**X**CACACACACACACACACA**Y**
XGTGTGTGTGTGTGT**Y**	**X**GTGTGTGTGTGTGTGTGT**Y**	**X**GTGTGTGTGTGTGTGTGT**Y**
XACACACACACACAC**Y**	**X**ACACACACACACAC**Y**	**X**ACACACACACACACACAC**Y**
XTGTGTGTGTGTGTG**Y**	**X**TGTGTGTGTGTGTGT**Y**	**X**TGTGTGTGTGTGTGTGTG**Y**

Fragment sizes on a gel (The samples loaded at top, migration is down the page, with smaller fragments coded for by the A_1 allele migrating furthest)

Microsatellite diversity is detected by amplifying DNA using PCR. Unique conserved sequences (primers) flanking microsatellites are used to specify the DNA region that is to be amplified. The resulting DNA fragments are separated according to size using electrophoresis on acrylamide or agarose gels. After separation the fragments are detected by either (1) staining gels with ethidium bromide (a DNA stain), (2) use of radioactively labelled primers and autoradiography of gels, or (3) use of fluorescently labelled primers and running the PCR products on a DNA sequencing machine. If an individual is heterozygous for two microsatellite alleles with different numbers of repeats, then two different-sized bands will be detected, as shown above.

Microsatellites typically measure genetic variation for loci that are neutral (not exposed to selection) since the tandem repeats are usually located in non-coding segments of the DNA.

DNA FINGERPRINTS: MINISATELLITES, OR VARIABLE NUMBER TANDEM REPEATS (VNTR)

Variable number tandem repeat sequences are found throughout the genome of humans and other eukaryotes. These minisatellite sequences have core repeat sequences with lengths in the range of 10–100 bases (i.e. they are larger than microsatellites). Typing of individuals for DNA fingerprints results in a 'barcode', where each individual is usually unique. To identify minisatellites, DNA is purified, cut with a restriction enzyme that cleaves outside the repeat, releasing the mini-satellite DNA fragment, and the fragmented DNA separated according to size on an agarose gel. The two strands of the DNA fragments are separated (denatured) and transferred to a membrane (Southern blotting). The membrane with attached DNA is placed in a solution containing many copies of single-stranded radioactively labelled DNA of the core repeat sequence (probed). Radioactively labelled core sequences attach (hybridize) to minisatellite fragments on the membrane by complementary base pairing. Single-stranded unhybridized probe DNA is washed away, the membrane is dried and the position of minisatellites is revealed by autoradiography.

The number of repeats is highly variable, such that each individual in outbreeding species normally has a unique DNA fingerprint (apart from identical twins). Three genotypes for a single minisatellite locus are illustrated below, along with their banding patterns on a gel. 'o' represents a single repeat of the core sequence. Many such loci are typed simultaneously, resulting in a pattern of bands akin to a barcode.

-----ooooo-----   -----oooooo-----   -----oooooo-----
-----ooooo-----   -----ooooo-----   -----oooooo-----

DNA fragments on a gel

DNA fingerprints are highly variable, assess nuclear DNA variation over a wide range of loci, and do not require prior knowledge of DNA sequence in the species being typed. The disadvantages are that individual loci are not normally identifiable, as the fragments derive from many different places in the genome. The inheritance of bands is not defined and they cannot be typed following non-invasive sampling, as they require considerable amounts of DNA. DNA fingerprints are now being replaced by methods that allow non-invasive sampling followed by PCR, such as microsatellites, AFLP or RAPDs.

DNA SEQUENCING

The most direct means for measuring genetic diversity is to determine the sequences of bases in the DNA. This is usually done using DNA sequencing machines. It is still relatively time-consuming and expensive, and has been used primarily for taxonomic purposes, where mtDNA and/or nuclear loci are sequenced for a small number of individuals. However, technical improvements have markedly reduced the cost and time taken to sequence DNA, as is evident from the Human Genome Project and similar endeavours for other species.

OTHER METHODS

Other methods including restriction fragment length polymorphisms (RFLP), random amplified polymorphic DNA (RAPD), amplified fragment length polymorphisms (AFLP), single-strand conformational polymorphisms (SSCP) and single nucleotide polymorphisms (SNP) are detailed in Frankham et al. (2002).

Mitochondrial DNA (mtDNA)

Mitochondria contain small circular DNA molecules that are maternally inherited (mother to offspring) in most species. **mtDNA** is relatively abundant, as there are many mitochondria per cell, and easy to purify. Genetic diversity in this DNA can be detected by a range of methods, including cutting with restriction enzymes (RFLP), SSCP and sequencing (Box 2.2). DNA primers that work for most species are now available for several loci in mtDNA. These loci can be amplified by PCR and the products sequenced. Sequencing of mtDNA has the advantages over other techniques that it can be done following non-invasive sampling, that mtDNA has a high mutation rate and is highly variable, and that it can be used to specifically trace female lines of descent, or migration patterns. Its disadvantages are that it traces only a single maternally inherited unit. We will defer further consideration of mtDNA variation to Chapters 6 and 9, as its main conservation uses are in resolving taxonomic uncertainties, defining management units, and in helping understand important aspects of species biology.

The organellar DNA in the chloroplasts of plants can be used for similar purposes.

> Mitochondrial DNA is maternally inherited in most species. It is used widely to assess taxonomic relationships and differences among populations within species

Levels of genetic diversity in DNA

The first extensive study of DNA sequence variation at a locus within a population was for the alcohol dehydrogenase (*Adh*) locus in fruit flies. Among 11 samples, there were 43 polymorphic sites across 2379 base pairs. The majority of base changes (42/43) did not result in amino acid substitutions (i.e. they are **silent**, or **synonymous substitutions**) as they were in non-coding regions of the locus (**introns**), or involved the third position in triplets coding for amino acids. Such variation is not detectable by protein electrophoresis. Sequence variation at the

> There is extensive genetic diversity in DNA sequences among individuals within outbreeding species

Most polymorphisms at the DNA level have little functional significance, as they occur in non-coding regions of the genome, or do not alter the amino acid sequence of a protein

Adh locus in two outbreeding plant species from the Brassica family was, if anything, even higher than that found in fruit flies.

The highest levels of genetic diversity in DNA are typically found in sequences with little functional significance. These variants either do not code for functional products, or substitutions do not change the function of the molecule, i.e. they are not expressed phenotypically. Therefore, selection does not act against such changes. Conversely, the lowest genetic diversity is found for functionally important regions of molecules, such as the active sites of enzymes. Much of the DNA in an organism does not code for functional products.

There are two important exceptions to the generalization that polymorphism is lowest in regions with important functions. These are the major histocompatibility complex (MHC – a large family of genes that play a central role in the vertebrate immune system and in fighting disease) and self-incompatibility (SI) loci in plants. Both regions have extremely high levels of genetic diversity due to natural selection that favours differences among individuals within populations.

Microsatellites provide one of the most powerful and practical means currently available for surveying genetic diversity in threatened species

Microsatellites are now used routinely to measure genetic diversity in a variety of species, many of them endangered. They typically show very high levels of polymorphism and many alleles per locus. For example, CA repeats are common and often vary in repeat number within a population. Such diversity has been found in all species so far examined. Data from a survey of microsatellite variation at eight loci in 39 wild chimpanzees detected an average of 5.75 alleles

Komodo Mago

Table 2.6 Microsatellite variation at eight loci in 39 wild chimpanzees from Gombe National Park, Tanzania. Allele frequencies at each of the eight loci are given, along with the proportions of individuals heterozygous (*H*) for each locus and average number of alleles per locus (*A*).

Allele	\<-- Locus --\>							
	D19S431	D9S905	D18S536	D4S243	D1S548	D9S922	D2S1326	D9S302
1	0.458	0.086	0.412	0.197	0.219	0.016	0.258	0.071
2	0.097	0.186	0.074	0.224	0.516	0.210	0.182	0.014
3	0.042	0.057	0.074	0.210	0.266	0.064	0.061	0.057
4	0.014	0.243	0.265	0.013		0.290	0.288	0.014
5	0.028	0.429	0.176	0.132		0.355	0.182	0.071
6	0.361			0.224		0.064	0.030	0.100
7								0.586
8								0.029
9								0.057
H	0.647	0.712	0.718	0.799	0.615	0.737	0.780	0.629

Av. *H* = 0.705

A = 5.75

Source: Constable *et al.* (2001).

per locus and an average heterozygosity of 0.70 (Table 2.6). The resolving power of microsatellites enables the determination of paternities and identification of migrants, a resolution usually not possible with allozymes.

Komodo dragon

Low genetic diversity in threatened species

The abundant genetic diversity found in large populations contrasts with that found in many small or bottlenecked populations. For example, the northern elephant seal has been hunted almost to extinction, but has subsequently recovered from its population size bottleneck. This species exhibits no allozyme variation and possesses substantially reduced levels of microsatellite variation in comparison to related non-bottlenecked species.

Small or bottlenecked populations often have reduced genetic diversity

Most threatened species have lower genetic diversity than related non-endangered species with large population sizes (Table 2.7). Of 170 threatened species examined, 77% had lower genetic diversity than related non-endangered species. For example, endangered species such as the northern hairy-nosed wombat and the Ethiopian wolf have greatly reduced microsatellite variation compared to related non-endangered species. Overall, threatened species have about 60% of the genetic diversity of non-endangered species. As we shall see,

Threatened species usually have lower levels of genetic diversity than non-endangered species

Table 2.7 Levels of microsatellite genetic diversity in a range of threatened, and related non-endangered species. Average number of alleles per locus (*A*) and heterozygosity (*H*) are given for polymorphic loci. Globally threatened, or previously threatened, species are placed adjacent to the most closely related, but non-endangered species for which data are available

Threatened species	*A*	*H*	Non-endangered species	*A*	*H*
Black rhinoceros	4.2	0.69	African buffalo	8.6	0.73
Mexican wolf	2.7	0.42	Gray wolf	4.5	0.62
Ethiopian wolf	2.4	0.21	Coyote	5.9	0.68
African wild dog	3.5	0.56	Domestic dog	6.4	0.73
Cheetah	3.4	0.39	African lion	4.3	0.66
Mariana crow	1.8	0.16	American crow	6.0	0.68
Mauritius kestrel	1.4	0.10	European kestrel	5.5	0.68
Seychelles kestrel	1.3	0.12	Greater kestrel	4.5	0.59
Peregrine falcon	4.1	0.48	Lesser kestrel	5.4	0.70
			Koala	8.0	0.81
Northern hairy-nosed wombat	2.1	0.32	Southern hairy-nosed wombat	5.9	0.71
Long-footed potoroo	3.7	0.56	Allied rock wallaby	12.0	0.86
Bridled nail-tail wallaby	11.6	0.83			
Komodo dragon	4.0	0.31	American alligator	8.3	0.67
Mahogany tree	9.7	0.55	Royal mahogany	9.3	0.67

Source: See Frankham *et al.* (2002).

the most likely explanation is that threatened species have suffered reductions in population size that directly result in loss of genetic diversity.

What components of genetic diversity determine the ability to evolve?

Quantitative genetic variation for life history traits is the major determinant of evolutionary potential. Unfortunately, we have least information on this form of genetic diversity and it is the most difficult to measure

Cheetah

Evolutionary potential is most directly measured by estimating the quantitative genetic variation for reproductive fitness. Reproductive fitness is the number of fertile offspring contributed to the next generation by an individual and encompasses survival to sexual maturity, and the ability to mate and raise offspring. Unfortunately, this quantitative genetic variation is the most difficult to measure and is the aspect of genetic diversity for which we have least information in threatened species. Other measures such as DNA and allozymes only reflect evolutionary potential if they are correlated with quantitative genetic variation. Disturbingly, correlations between molecular and quantitative measures of genetic diversity are often low.

SUGGESTED FURTHER READING

Frankham, R., J. D. Ballou & D. A. Briscoe. 2002. *Introduction to Conservation Genetics*. Cambridge University Press, Cambridge, UK. Chapters 3–5 have extended treatments of these topics, plus references.

Avise, J. C. & J. Hamrick. (eds.) 1996. *Conservation Genetics: Case Histories from Nature*. Chapman & Hall, New York. Advanced scientific reviews on the conservation of major groups of animals and plants, including considerable information on genetic diversity.

Smith, T. B. & R. K. Wayne. (eds.) 1996. *Molecular Genetic Approaches in Conservation*. Oxford University Press, New York. Contains information on genetic diversity in threatened species and the diverse molecular methods for measuring it. Advanced.

Chapter 3

Evolutionary genetics of natural populations

Populations evolve through the actions of mutation, migration, selection and chance. Genetic diversity arises from mutation, or is added by immigration, and is removed by selection, or lost by sampling in small populations. The balance between mutation and selection results in a 'load' of rare deleterious alleles

Terms

Adaptive evolution, additive variance, cline, dominance variance, evolution, fitness, gene flow, genetic marker, genotype × environment interaction, heritability, interaction variance, introgression, lethal, mutation, mutational load, mutation–selection balance, natural selection, neutral mutation, relative fitness, reproductive fitness, selection coefficient, selectively neutral

Industrial melanism in the peppered moth; peppered and melanic (black) moths on trees in polluted (blackened tree trunk) and unpolluted (tree with lichen) areas of England. The melanic form is better camouflaged in the polluted area and the peppered form in the unpolluted area (from Kettlewell 1973).

Factors controlling the evolution of populations

Populations evolve through the action of selection, mutation, migration, and chance

Our objective in conservation genetics is to preserve species as dynamic entities, capable of evolving to adapt with environmental changes. It is therefore essential to understand the natural forces determining evolutionary change. This information is central to planning genetic management of threatened and endangered populations. Since **evolution**, at its most basic level, is change in the genetic composition of a population, it requires genetic diversity. Consequently, we must appreciate how genetic diversity arises, what forms of genetic diversity exist and how it is lost.

Evolution at its simplest level is a change in the frequency of an allele

At its simplest level, evolution involves any change in the frequency of an allele due to mutation, migration (**gene flow**), selection or chance.

The roles of these factors can be summarized as follows:

- **mutation** is the source of all genetic diversity, but is a weak evolutionary force over the short term, as mutation rates are generally very low
- **migration** (gene flow) reduces differences among populations generated by mutation, selection and chance
- **selection** is the only force causing evolutionary changes that better adapt populations to their environment
- **chance** effects in small populations lead to loss of genetic diversity
- **fragmentation** and reduced migration limiting gene flow generate random differentiation among subpopulations derived from the same original source population.

An evolving population can be modelled as a complex system influenced by mutation, migration, selection and chance, operating within the context of the breeding system (Fig. 3.1). To evaluate the importance of the components of an evolving population, we model it with none of the factors operating, then with each of the factors

Fig. 3.1 An evolving population as a complex system.

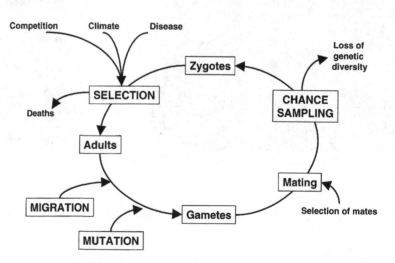

individually, followed by two at a time, etc. By so doing we can estimate the impact of each factor and what role it is likely to play in evolution. Further, we can identify those circumstances where particular factors can be ignored, as it is rare for all factors to have significant effects simultaneously. For example, mutations occur at very low rates, and we can often ignore them over the short term.

In the previous chapter we showed that allele and genotype frequencies at an autosomal locus are in Hardy–Weinberg equilibrium after one generation of random mating in populations free from mutation, migration, selection or chance. Below we consider the independent action of mutation, migration and selection, and then the joint actions of mutation with selection. Chance effects are generally minor in large populations, and we defer detailed treatment of them until Chapter 4.

Origin and regeneration of genetic diversity

Genetic diversity is the raw material upon which natural selection acts to bring about **adaptive evolutionary change**. Most naturally outbreeding species with large populations carry a substantial store of genetic diversity (Chapter 2). Here we address the questions how is genetic diversity produced, and how quickly is it regenerated if it is lost?

Mutation

A **mutation** is a sudden genetic change in an allele or chromosome. All genetic diversity originates from mutation. The word mutation refers both to the process by which novel genetic variants arise (through natural errors in DNA replication, mobile genetic elements, chromosomal breakages, etc.) and to the products of the mutational process (e.g. the white eye mutation in fruit flies).

Mutation is the ultimate source of all genetic diversity

The patterns of genetic diversity in populations are the result of a variety of forces that act to eliminate or increase and disperse these novel mutant alleles and chromosome arrangements among individuals and populations. In conservation genetics, we are concerned with:

- How rapidly does the process of mutation add genetic diversity to populations?
- How do mutations affect the adaptive potential and reproductive fitness of populations?
- How important is the accumulation of deleterious mutations to fitness decline in small populations?

The most important mutations are those at loci affecting fitness traits, most notably lethal or deleterious mutations. Some mutations, such as the dark melanic allele in moths, actually increase fitness in polluted areas, as moths carrying the mutation are better camouflaged on blackened trees and so suffer less bird predation. However, many mutations that occur in non-coding regions of the genome and those that do not result in amino acid substitutions in proteins (**silent substitutions**) probably have little or no impact on fitness (**neutral mutations**). Neutral mutations are, however, important as molecular markers and clocks that provide valuable information on genetic differences among individuals, populations and species.

The rate of mutation is critical to its role in evolution. Rates are low. For a range of loci in eukaryotic species, the typical spontaneous mutation rate is one new mutation per locus per 100 000 gametes (10^{-5}) per generation (Table 3.1). Mutation rates are similar across all eukaryotes, apart from those for microsatellites.

Mutation rates per nucleotide base are clearly lower than for single loci as there are typically 1000 or more bases per gene locus. Mitochondrial DNA has a much higher mutation rate than nuclear loci, making it a valuable tool in studying short-term evolutionary processes. Mutation rates for quantitative characters are approximately 10^{-3} times the environmental variance per generation for a range of characters across a range of species. This apparently high rate, compared to single loci, is because a mutation at any of the many loci underlying the character can affect the trait.

Mutation is a recurrent process where new alleles continue to arise over time. It is in fact a slow chemical reaction. We can model the impact of mutation on a population by considering a single locus

> Mutations typically occur at a very low rate

> As mutation rates are very low, mutation is a very weak evolutionary force and can be ignored in the short term for many circumstances

Table 3.1 | Spontaneous mutation rates for different loci and characters in a variety of eukaryote species. Approximate mean rates are given as the frequency of new mutations per locus per generation, except where specified otherwise

Morphological mutations	
Mice, maize and fruit flies (normal $\rightarrow$ mutant)	$\sim 1 \times 10^{-5}$/locus
Allozyme loci (mobility change)	0.1×10^{-5}/locus
Microsatellites Mammals	1×10^{-4}/locus
DNA nucleotides	$10^{-8} - 10^{-9}$/base
mtDNA nucleotides Mammals	$5 - 10 \times$ nuclear
Quantitative characters[a] Fruit flies, mice and maize	$10^{-3} \times V_E$/trait

[a] V_E is the environmental variation for the character.
Sources: Houle *et al.* (1996); Hedrick (2000).

with two alleles A_1 and A_2 at frequencies of p and q, with mutations only changing A_1 into A_2 at a rate of u per generation, as follows:

$$\text{Mutation rate}$$
$$u$$
$$A_1 \longrightarrow A_2$$
$$\text{Initial allele frequencies} \quad p_0 \qquad q_0$$

The frequency of the A_1 allele in the next generation p_1 is the frequency of alleles that do not mutate, namely:

$$p_1 = p_0(1 - u) \tag{3.1}$$

Thus, the frequency of the A_1 allele declines.

The change in frequency of the A_1 allele (Δp) is the difference between the frequencies in the two generations:

$$\Delta p = p_1 - p_0 = p_0(1 - u) - p_0$$
$$\Delta p = -up_0 \tag{3.2}$$

Consequently, the frequency of A_1 declines by an amount that depends on the mutation rate u and the starting frequency p_0. There is a corresponding increase in the frequency of A_2 ($\Delta q = +up_0$). Since the mutation rate is approximately 10^{-5} for morphological mutations, the maximum change in allele frequency is 10^{-5} when $p = 1$. This is very small and can be ignored in many circumstances.

When genetic diversity is lost from an entire species, it is only regenerated by mutation. As mutation rates are low, regeneration times are very long, typically taking thousands to millions of generations for single locus variation (Chapter 5).

> Genetic diversity is only regenerated by mutation over periods of hundreds to millions of generations

Migration and gene flow

Unlike the slow process of mutation, mixture of alleles from two or more genetically differentiated populations can rapidly restore genetic diversity. The benefits, and potential hazards, of managed gene mixing to restore genetic diversity in endangered species are discussed in Chapter 7.

> The introduction of immigrants from one population into another reduces genetic differentiation among populations and may restore lost genetic diversity

The gene pools of partially isolated populations diverge over time as a result of chance and selection. Migration and subsequent interbreeding reduce such differences. The impact of migration is illustrated by B blood group allele frequencies in human populations across Eurasia (Fig. 3.2). Prior to about 1500 years ago, the B allele was essentially absent from Western Europe, but it existed in high frequencies in the east. However, between 1500 and 500 years ago, there was a succession of Mongol and Tartar invasions of Europe. As was typical of such military invasions, they left a trail of pillage and rape, and consequently left some eastern alleles behind. Note the gradual decrease in frequency of the B blood group allele from east to west (termed a **cline**).

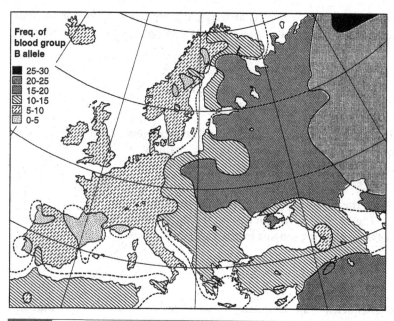

Fig. 3.2 B blood group allele frequencies across Eurasia, resulting from the Mongol and Tartar invasions between 1500 and 500 years ago (from Mourant *et al.* 1976). Prior to this, the B blood group allele was presumed to be absent from western Europeans, as it is absent in native Basques in Spain, and in other isolated populations.

The genetic impact of migration depends on the proportion of alleles contributed by immigrants and on the difference in frequency between the native population and the immigrants

The change in allele frequency due to migration is

$$\Delta q = m(q_m - q_o) \tag{3.3}$$

where m is the migration rate, q_m the allele frequency in immigrants and q_o the frequency in the original population.

Thus, the change in allele frequency from one generation to another depends on the proportion of alleles contributed by immigrants, and on the difference in frequency between the immigrants and the original population. Migration may have very large effects on allele frequencies and is much more effective at restoring genetic diversity than mutation. For example, if immigrants are homozygous for an allele absent from the native population, and 20% of the population in the next generation are immigrants, then the immigrant allele increases in frequency from 0 to 0.2 in a single generation. In Chapter 4 we show how migration rates as low as one migrant per generation are enough to keep populations from diverging genetically.

Many endangered species are threatened by gene flow (**introgression**) from related non-endangered species. Equation 3.3 can be used to estimate the immigration rate from allele frequency data. For example, approximately 22% of the genetic material in the Web Valley population of endangered Ethiopian wolves derives from domestic dogs (Example 3.1).

Example 3.1	Estimating dog introgression in the endangered Ethiopian wolf from microsatellite allele frequencies (data from Gottelli et al. 1994)

Ethiopian wolves are genetically distinct from domestic dogs, but hybridization occurs in areas where they co-occur, as in Web Valley, Ethiopia. The population from the Sanetti Plateau is relatively pure. The contribution of genetic material from domestic dogs in the Web Valley population can be estimated using the allele frequencies at microsatellite locus 344. Dogs lack the J allele, while 'pure' Ethiopian wolves are homozygous for it. Frequencies of this allele are:

Population		J allele frequency
Sanetti	q_0	1.00 ('old' – 'pure' Ethiopian wolf)
Web Valley	q_1	0.78 ('new' – containing dog admixture)
Domestic dogs	q_m	0.00 ('immigrants')

All the non-J alleles in the Web Valley population have come from dogs. Equation 3.3 can be rearranged to provide an expression for the migration or introgression rate m

$$\Delta q = q_1 - q_0 = m(q_m - q_0)$$

Thus,

$$m = \frac{q_1 - q_0}{q_m - q_0}$$

Upon substituting allele frequencies from above into this expression, we obtain

$$m = \frac{0.78 - 1.0}{0 - 1.0} = 0.22$$

Thus, the Web Valley population of Ethiopian wolves derives about 22% of its genetic composition from domestic dogs. This is the accumulated contribution of alleles from dogs, not a per-generation estimate. Phenotypically abnormal individuals, suspected of being hybrid individuals, represent about 17% of the population. Estimates can also be made from other microsatellite loci and the best estimate would come from combining information from all relevant loci.

Selection and adaptation

The physical and biotic environments of virtually all species are continually changing. For long-term survival, species must adjust to these changes. Climates fluctuate over time, sea levels make and break connections between landmasses, and ice caps advance and retreat. Pests,

Species evolve in response to environmental change. Adaptive evolutionary changes occur through the impact of selection on genetic variation, increasing the frequency of beneficial alleles parasites and diseases evolve new strains, switch hosts and spread to new locations, and new competitors may arise. Adaptive evolution has been described in a large number of animals and plants. For example, evolutionary changes have been documented in animals for morphology, behaviour, colour form, host plant resistance, prey size, body size, alcohol tolerance, reproductive rate, survival, disease resistance, predator avoidance, tolerance to pollutants, biocide resistance, etc. Adaptive evolutionary changes to a wide range of conditions have been reported in plants, including those to soil conditions, water stress, flooding, light regimes, exposure to wind, grazing, air pollution and herbicides.

Of particular concern during the 'sixth extinction' are the environmental changes wrought by human activities. Global warming is occurring as a consequence of industrial and agricultural practices. Species have to move, or adapt to these changed climatic conditions. Birds, butterflies and plants have already altered their ranges. Humans have also been responsible for translocations of species, extinction of food species, and inadvertent or deliberate introduction of novel chemicals to the environment. In many cases, adaptive evolutionary changes have been recorded in affected species. Rabbits in Australia rapidly evolved resistance to the myxoma virus when it was introduced as a control measure (Box 3.1). Over 200 species of moths worldwide display melanism in polluted industrial areas. Several species of plants have evolved tolerance to heavy metals in the process of colonizing polluted heavy-metal mine wastes and plants are progressively evolving resistance to herbicides.

Box 3.1	Rapid adaptive evolutionary changes in rabbits in Australia following the introduction of myxoma virus as a control agent (Fenner & Ratcliffe 1965)

European settlers introduced rabbits into Australia in the nineteenth century for sport hunting. Several unsuccessful attempts were made until genuine wild rabbits were introduced in 1859. The wild rabbits rapidly increased in numbers until they reached plague proportions throughout much of the country. Rabbits caused many native plant species to decline and were one of the causes in the decline of native marsupial bilbies which also burrow.

When myxoma virus was introduced into Australia to control rabbits in 1950, mortality rates of infected rabbits were over 99%. Strong directional selection resulted in rapid increases in genetic resistance of rabbits to the myxoma virus. The myxoma virus also evolved lower virulence, as this increased the probability of being transmitted. The data below reflects only the genetic changes in rabbit resistance as the same less severe virus stain was used throughout the study. The mortality to this virus strain dropped from around 90% to 25% in 1958 (the sixth epizootic).

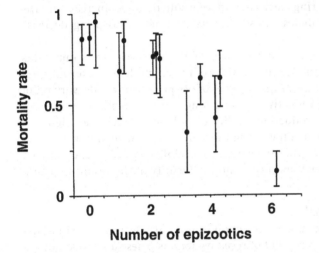

Adaptation may take the form of either physiological or behavioural modifications where individuals change to cope with altered conditions, or genetic adaptation where natural selection alters the genetic composition of populations over several or many generations.

Physiological adaptations by individuals include modifications in haemoglobin levels to cope with altitude, immune responses to fight diseases, induction of enzymes to cope with altered diets, etc. Behavioural adaptations may include altered food preference, avoidance behaviours, etc. There is however a limit to physiological adaptation. If environment changes are so extreme that no individual can cope, then the species becomes locally or globally extinct.

Evolutionary change through **natural selection** is the long-term mechanism for coping with environmental change. This is referred to as **adaptive evolution**. When adaptive evolutionary changes continue over time, they may allow a population or species to thrive in conditions more extreme than any individual could originally tolerate.

In the face of environmental change, species must adapt, or face extinction

Adaptive evolutionary changes may allow populations to cope with conditions that no individual could previously survive

Adaptive evolution is observed wherever large genetically variable populations are subjected to altered biotic or physical environments. It is of major importance in five conservation contexts:

- preservation of the ability of species to evolve
- loss of adaptive evolutionary potential in small populations
- most endangered species now exist only on the periphery of their historical range, and must therefore adapt to what was previously a marginal environment
- genetic adaptation to captivity and its deleterious effects on reintroduction success (Chapter 8)
- adaptation of translocated populations to their new environment

Selection arises because different genotypes have different rates of survival and reproduction, resulting in changes in the frequency of alleles. Alleles whose carriers produce relatively larger numbers of fertile offspring surviving to reproductive age increase in frequency, while alleles whose carriers have fewer offspring decrease in frequency.

Selection operates at all stages of the life cycle. In animals this involves mating ability and fertility of males and females, fertilisation success of sperm and eggs, number of offspring per female, survival of offspring to reproductive age and longevity. In plants, selection can involve pollen production, ability of pollen to reach the stigma of flowers, germinate, grow down the style and fertilize, number of ova, viability of the fertilized zygotes, their ability to disperse, germinate and grow to sexual maturity, and the fertility of the resulting plant.

Selection is the only force that causes adaptive evolution

Recessive lethal

Selection reduces the frequency of deleterious alleles

To illustrate and model the action of natural selection we will examine an extreme case, that of recessive lethals. These alleles do not impair the ability of heterozygotes but all homozygotes die (**lethal**). For example, all individuals homozygous for chondrodystrophic dwarfism (*dwdw*) in endangered California condors die around hatching time. The effect of selection against chondrodystrophy is modelled in Box 3.2. We begin with a normal allele (+) at a frequency of p and the recessive lethal (*dw*) allele at a frequency of q. With random mating, the genotype frequencies at zygote formation are the Hardy–Weinberg equilibrium frequencies p^2, $2pq$ and q^2. However, the three genotypes have different survival. The important factor in the genetic effects of selection is not the absolute survival, but the relative survival of the three genotypes. For example, if the ++, +*dw* and *dwdw* genotypes have 75%, 75% and 0% survival, the relative values 1, 1 and 0 determine the impact of selection. We term these values the **relative fitnesses**, where the genotype(s) with the highest fitness are given the maximum value of 1.

The frequency of surviving adults is obtained by multiplying the initial frequencies by the relative fitnesses. For example, the frequency of lethal homozygotes goes from q^2 at fertilization to $q^2 \times 0 = 0$ in adults. After selection, we have lost a proportion of the population ($-q^2$), so the total no longer adds to 1. We must therefore divide by the total ($1 - q^2$) to obtain relative frequencies, as shown.

The allele frequency in the succeeding generation is then obtained by determining the allele frequency in survivors using the allele counting method described previously. Methods for predicting allele frequency changes due to any form of selection are similar to that used here.

Box 3.2 Modelling the impact of selection against chondrodystrophy (a recessive lethal) in California condors

	Normal	Normal	dwarf	
	++	+dw	dwdw	Total
Zygotic frequencies	p^2	$2pq$	q^2	1.0
Relative fitnesses	1	1	0 (lethal)	
After selection (frequency × fitness)	$p^2 \times 1$	$2pq \times 1$	$q^2 \times 0 = 0$	$1 - q^2$
Adjusted frequencies	$\frac{p^2}{1-q^2}$	$\frac{2pq}{1-q^2}$	0	1.0

(Column headings: Phenotype — Normal, Normal, dwarf; Genotype — ++, +dw, dwdw.)

The frequency of the *dw* allele in the next generation after selection (q_1) is

$$q_1 = \frac{\text{homozygotes} + \frac{1}{2}\text{ heterozygotes}}{\text{total}} = \frac{0 + pq}{1 - q^2} = \frac{q(1-q)}{(1-q)(1+q)}$$

$$q_1 = \frac{q}{1+q} \qquad (3.4)$$

Note that we substituted $1 - q$ for p, as $p + q = 1$.

The change in frequency Δq is:

$$\Delta q = q_1 - q = \frac{q}{1+q} - q = \frac{q - q(1+q)}{1+q}$$

$$\Delta q = \frac{-q^2}{1+q} \qquad (3.5)$$

California condor

Equation 3.5 demonstrates that the lethal *dw* allele always declines in frequency, as the sign of Δq is negative. Further, the rate of decline slows markedly at lower allele frequencies, as it is dependent on q^2. For example, if q is 0.5, Δq is -0.167, while if q is 0.1, Δq is -0.009. Example 3.2 uses equation 3.4 to calculate the change in frequency of the *dw* alleles in California condors.

> **Example 3.2** | How rapidly does the frequency of the recessive lethal chondrodystrophy allele decline due to selection in endangered California condors?
>
> The chondrodystrophy allele had an initial frequency at fertilization of about 0.17 (Chapter 2). All homozygotes die, thus the frequency is reduced in surviving adults. The expected frequency of the deleterious allele in adults as a result of this natural selection can be predicted by using equation 3.4 and substituting $q = 0.17$, as follows:
>
> $$q_1 = \frac{q}{1+q} = \frac{0.17}{(1+0.17)} = 0.145$$
>
> Thus, the frequency is predicted to drop from 17% to 14.5% as a result of one generation of natural selection.

Selection not only applies to lethals. Any allele that changes the relative fitness of its carriers will be subject to selection. If the effect on fitness is small then the change in allele frequency will be correspondingly smaller.

Selection increases the frequency of advantageous alleles

In conservation genetics, we are concerned both with selection against deleterious mutations (described above) and with selection favouring alleles that improve the ability of a population to adapt to changing environments. We will use industrial melanism in the peppered moth in Britain to illustrate adaptive evolutionary change. Camouflage is critical to avoidance of predation by birds as peppered moths are active at night and rest on trees during the day. Prior to the industrial revolution, its mottled wings were well camouflaged as it rested on speckled lichen-covered tree trunks in the midlands of England (chapter frontispiece). However, sulphur pollution following industrialization killed most lichen and soot darkened trees. The speckled moth became clearly visible. The previously rare dark variants (melanics) were better camouflaged on the black trunks and suffered less predation. This resulted in a higher frequency of the melanic allele (M) in industrial areas, than in relatively unpolluted areas (Box 3.3). The melanic form of the peppered moth was first recorded in 1848, but by 1900 they represented about 99% of all moths in polluted areas.

A model of this selection is developed in Box 3.3 and shows that the frequency of the M allele always increases until the allele is fixed ($p = 1$), as the sign of Δp is positive. The rate of change depends on the strength of selection against the non-melanic form (s, the **selection coefficient**) and the allele frequencies p and q.

Pollution controls would be expected to reverse the selective forces. As predicted, the frequency of melanics (now poorly camouflaged) declined markedly. At one site near Liverpool, it has dropped from 90% to 10% over the last 40 years, and similar declines have been observed in other areas of the UK. Parallel changes have also occurred in the North American sub-species of the peppered moth.

| **Box 3.3** | Adaptive changes in the frequency of industrial melanism due to selection in polluted areas (after Kettlewell 1973; Majerus 1998; Grant 1999) |

The map of the UK with pie diagrams shows the frequency of the melanic, a milder melanic (insularia) and the non-melanic (typical) forms of the peppered moths in the 1950s. The melanic form had high frequencies in industrial areas (Midlands, around London in the southeast and around Glasgow towards the northwest) and low frequencies in less polluted areas.

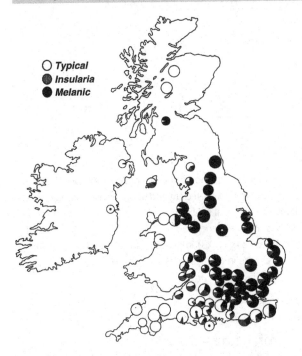

○ **Typical**
◑ **Insularia**
● **Melanic**

The melanic form of the moth is due to a single dominant allele. The impacts of selection on the melanic and typical allele frequencies are given below (the insularia allele is ignored here, but this does not affect our conclusions). We begin with frequencies for melanic (M) and typical (t) alleles of p and q, respectively and assume that we are dealing with a large random-mating population with no migration or mutation. Selection is assumed to occur on adults, but before reproduction. Since the tt genotype has poorer survival than melanics in polluted areas, but we do not necessarily know the precise value, we give it a relative fitness of $1 - s$, where s is the selection coefficient. The value of s represents the reduction in fitness of the tt genotype compared to the fittest genotypes, MM and Mt. For example, if the survival of tt was only 70% that of MM, then the selection coefficient is $1 - 0.7 = 0.3$.

	Melanic MM	Melanic Mt	Typical tt	Total
Zygotic frequencies	p^2	$2pq$	q^2	1.0
Relative fitnesses	1	1	$1 - s$	
After selection	p^2	$2pq$	$q^2(1 - s)$	$1 - sq^2$
Adjusted frequency	$\dfrac{p^2}{1 - sq^2}$	$\dfrac{2pq}{1 - sq^2}$	$\dfrac{q^2 - sq^2}{1 - sq^2}$	1.0

Frequency of M after selection (p_1) is

$$p_1 = \frac{p^2}{1 - sq^2} + \frac{(1/2)2pq}{1 - sq^2}$$

$$= \frac{p^2 + pq}{1 - sq^2} = \frac{p(p + q)}{1 - sq^2}$$

$$= \frac{p}{1 - sq^2}$$

The change in frequency of M (Δp) is

$$\Delta p = p_1 - p$$

$$= \frac{p}{1 - sq^2} - p = \frac{p - p(1 - sq^2)}{1 - sq^2}$$

$$= \frac{spq^2}{1 - sq^2}$$

Thus, the melanic allele increases in frequency, as the sign of Δp is positive. The rate of increase depends upon the selection coefficient and upon the allele frequencies. This can be illustrated using a numerical example.

If the melanic allele was at a frequency p of 0.005 in 1848, and typicals had only 70% the survival of melanics ($s = 0.3$) in polluted areas, then the frequency of the melanic allele would change in one generation to

$$p_1 = \frac{p}{1 - sq^2} = \frac{0.005}{1 - (0.3 \times 0.995^2)} = 0.0071$$

The change in frequency is

$$\Delta p = p_1 - p = 0.0071 - 0.005 = 0.0021$$

Thus, the melanic allele increased by 0.0021, from 0.005 to 0.0071 in the first generation, an increase of ~40%.

Selection on quantitative characters

The discussion above applies to selection on differences due to single loci. However, in conservation genetics we are concerned primarily with the evolution of reproductive fitness, a quantitative trait affected by many loci, and how the ability to adapt is affected by reduced population size, fragmentation, and changes in the environment. The immediate evolutionary potential of a population is determined by the **heritability**. We now explore this parameter in more detail.

Heritability (h^2) is a fundamental measure of how well a quantitative trait is transmitted from one generation to the next. Heritability

> The immediate evolutionary potential of a population is determined by the heritability

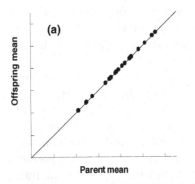
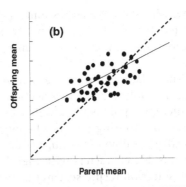
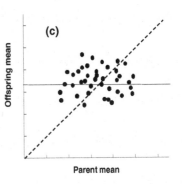

Fig. 3.3 Hypothetical relationships between mean values of parents and mean values of offspring for three cases, representing (a) complete, (b) incomplete and (c) zero relationships between parents and offspring. Solid lines are best fitting lines relating offspring and parents.

is the proportion of the total variation for a character in a population that is due to genetic differences, rather than environmental differences among individuals. It reflects the degree of resemblance between relatives for the character and it is most easily measured by comparing the trait among relatives.

Figure 3.3 illustrates three contrasting strengths of relationship between parents and offspring and the corresponding differences in heritabilities. Figure 3.3a shows an example of complete inheritance. Parents with larger than average values for the trait produce offspring with similar, larger, values, while smaller than average parents produce smaller than average offspring. This trait has a high heritability. In this case, the slope defining the relationship of parent mean to offspring mean (the regression coefficient) is 1. In this example, it is clear that environmental differences among parents, and among offspring, have negligible influence on the phenotype for the trait (a heritability of 1). An example that approaches this level of relationship is fingerprint ridge count in humans.

Inheritance is less complete in Fig. 3.3b. Parents with high phenotypic values produce offspring with higher than average values, but closer to the population mean than the parents themselves. Parents with low values produce offspring not as low as themselves. The slope of the relationship between offspring mean and parent mean is <1 and the heritability has an intermediate value. Some of the superiority or inferiority of the parents is due to environmental, rather than genetic effects and therefore cannot be inherited by their offspring. Many quantitative characters have relationships of this kind, including shell width in *Partula* snails and body size in many species, including endangered cotton-top tamarins.

There is no relationship between parent and offspring values in Fig. 3.3c. The slope of the relationship is 0. Parents with high and low values of the trait have similar offspring with values randomly distributed around the mean. In this case h^2 is zero. Such relationships are found in homozygous populations, such as the Wollemi

The slope of the relationship between offspring means and parent means is a direct measure of the heritability (h^2) of a trait

Partula snail

pine, where all differences among parents and among offspring are of environmental origin. Further, some reproductive characters in outbred populations, such as conception rate in cattle, approach this value.

The relationships in Fig. 3.3 also indicate how particular populations will respond to selection for particular characters. These predictions apply equally well whether we are considering natural selection favouring certain phenotypes, or when we are conducting selective breeding in the improvement of our domesticated plants and animals. For characters with high heritabilities (Fig. 3.3a) we predict high **selection response**. Parents displaying larger values for the trait do so because they carry alleles determining large values. If natural selection, or an agricultural geneticist, determines that only those individuals with values a certain level above the population mean are to become parents, then their offspring will inherit high-value alleles and all of the improvement. In contrast, where heritability is zero (Fig. 3.3c), all of the differences among potential parents are due to non-heritable environmental differences, it therefore does not matter which individuals reproduce, and the offspring generation will have a mean similar to that of the parental generation. Intermediate heritabilities (Fig. 3.3b) predict intermediate selection response. For example, a heritability of 0.5 predicts that offspring will have a mean that deviates from the population mean by half as much as their parents did. Heritabilities are direct predictors of evolutionary response.

As noted above, heritability is the proportion of total variance in a trait or character due to genetic diversity:

$$h^2 = \frac{V_G}{V_P} \tag{3.6}$$

where V_G is the proportion of variation due to genetic differences among individuals and V_P is the total amount of phenotypic variation. The amount of genetic variation for the trait is determined by the level of heterozygosity at the loci involved in determining the trait.

The total amount of variation in a character among all individuals (V_P) is composed of variation due to genetic differences among individuals (V_G) and differences caused by the environment (V_E):

$$V_P = V_G + V_E \tag{3.7}$$

V_E can be calculated by measuring the amount of variation in the character among individuals that are genetically identical (e.g. a population made completely homozygous through inbreeding). $V_P - V_E$ then gives V_G.

The genetic variation can be further divided into **additive, dominance** and **interaction variances**

$$V_G = V_A + V_D + V_I \tag{3.8}$$

V_A is the variation due to the average effects of alleles and is the component that determines immediate evolutionary potential. Strictly, V_A

> Heritability is directly related to the level of heterozygosity in a population

Table 3.2 Heritabilities of beak, body size (or tarsus length) and fitness characters for wild birds. Values greater than 100% or less than 0 can arise due to sampling variation in small experiments

Species	h^2 (%)		
	Fitness	Body size	Bill size
Canada goose		11	46
Collared flycatcher	−5, 0, 29, 32	47, 59	35, 48, 56, 40, 44
Darwin's medium ground finch	−17	42, 61, 95	75, 102, 103,108
Darwin's cactus finch		37, 110, 126	2, 13, 44, 129
Darwin's large cactus finch		54, 95	67, 69, 104, 137
European starling	34	49	
Great tit	37, 48	59, 59, 61, 64, 76	49, 71, 68
Penguin		92	76
Pigeon		28, 50	50, 58
Red grouse	30	35, 50	
Song sparrow		27, 36, 71, 101	40, 123, 71, 59
Mean	**21**	**61**	**67**

Sources: Smith (1993); Weigensberg & Roff (1996); Lynch & Walsh (1998).

is the component of genetic variation that enters into the heritability ($h^2 = V_A/V_P$) for prediction of selection response. V_D is due to variation in dominance and reflects susceptibility to inbreeding depression. V_I is due to interactions among loci and is an important determinant of the impact of outcrossing.

Heritabilities are specific to particular characters for particular populations living under particular environmental conditions. Different populations may have different levels of genetic variation; those with greater heterozygosity will have greater heritabilities when compared in the same environment. Despite these provisos, heritability estimates show relatively consistent patterns in magnitude for similar characters among populations within species, and across species (Table 3.2).

A heritability estimate is specific to a particular character in a particular population in a particular environment

A notable feature of heritabilities is that they are lower for fitness characters than for characters more peripherally related to fitness (Table 3.2).

Genotype × environment interaction

Differences in performance of genotypes in different environments are referred to as **genotype × environment interactions**. They typically develop when populations adapt to particular environmental conditions, and survive and reproduce better in their native conditions than in other environments. Genotype × environment interactions may take the form of altered rankings of performance in different environments or magnitudes of differences that vary in

Darwin's medium ground finches: differences in bill size produced by natural selection

Genotypes may show different performances in different environments diverse environments. A classical example is provided by the growth and survival of transplanted individuals of the sticky cinquefoil plant from high, medium and low elevations in California (Fig. 3.4). When grown in each of the three environments, strains generally grew best in the environment from which they originated and poorest in the most dissimilar environment.

Grown at sea level	Grown at 1400m	Grown at 3050m

Potentilla g. nevadensis from 3050m

Potentilla g. hanseni from 1400m

No survivors

Potentilla g. typica from sea level

Fig. 3.4 Genotype × environment interaction in the sticky cinquefoil plant. Strains of cinquefoil derived from high, medium and low altitudes were transplanted into their native and different locations in California and their growth and survival monitored (after Clausen et al. 1940). *Populations generally grow best in their own environment and poorest in the environment most dissimilar from their own.*

Genotype × environment interactions are of major significance to the genetic management of endangered species because:

- the reproductive fitness of translocated individuals cannot be predicted if there are significant, but poorly understood, genotype × environment interactions
- success of reintroduced populations may be compromised by genetic adaptation to captivity – superior genotypes under captive conditions may have low fitness when released to the wild
- mixing of genetic material from populations from different environments may generate genotypes that do not perform well under some, or all, conditions
- knowledge of genotype × environment interaction can strongly influence the choice of populations for return to the wild

Mutation–selection balance

Selective value of mutations

Earlier in this chapter we noted that new mutations are being continually added to populations, albeit at a very slow rate. We have also considered the evolutionary fate of highly deleterious alleles (recessive lethals) and beneficial alleles. What proportion of mutations belong in each category?

The majority of newly arisen mutations are deleterious

As the majority of the genome is not expressed in the phenotype (i.e. non-coding DNA), mutations in these regions will not affect fitness and be **selectively neutral**. These mutations are, however, invaluable in conservation genetics as they provide **genetic markers** (alleles that differ in frequency among individuals or populations) such as microsatellites for identifying individuals, populations and species. Mutations that alter the amino acid sequences of proteins, but do not affect the physiological functioning of the protein, are also selectively neutral and provide markers (e.g. allozymes).

Mutations within functional loci will be predominantly deleterious as random changes in the DNA sequence of a locus will usually be from a functional allele to a less-functional state. A small proportion of mutations is advantageous.

There is limited evidence on the proportions of mutations that fall into the different categories. As the majority of DNA is not involved in coding for proteins or other obvious functions, most mutations are probably neutral or near neutral. Of those with phenotypic effects, the great majority are deleterious and perhaps only 1–2% have beneficial effects.

Deleterious mutations occur at many loci, so their cumulative rates per gamete are much higher than the rates given above for single loci. These cumulative rates are of major importance when we consider the total load of mutations in populations and the impact of inbreeding on them. For example, the total rate of recessive lethal mutation is about 0.01 per haploid genome per generation in fruit

Deleterious mutations occur at thousands of loci in the genome, so their cumulative rate is relatively high flies, nematodes and ferns. As there are other deleterious mutations that are not lethal, the cumulative rate of deleterious mutations is considerably higher.

The mutation 'load'

Low frequencies of deleterious alleles are found at many loci in all outbreeding populations, due to the balance between their addition by mutation and their removal by natural selection

While selection is capable of removing deleterious alleles from populations, in reality the time taken is so long that new mutations usually occur before previous mutations have been eliminated, especially for recessive alleles. A balance (equilibrium) is reached between addition of deleterious alleles by mutation and their removal by selection (**mutation–selection balance**). Consequently, low frequencies of deleterious alleles are found in all naturally outbreeding populations (**mutation load**). These mutations are extremely important in understanding the deleterious consequences of inbreeding because inbreeding increases the probability of these alleles being expressed in homozygous genotypes (Chapters 4 and 5).

We noted earlier that population models can combine the effects of several processes. The equation for equilibrium frequency (q) due to mutation and selection acting simultaneously on a deleterious autosomal recessive allele is

$$\hat{q} = \sqrt{\frac{u}{s}} \tag{3.9}$$

where u is the mutation rate and s is the selection coefficient. For example, chondrodystrophy in California condors is a recessive lethal, so $s = 1$. If $u = 10^{-5}$, the equilibrium frequency is predicted to be $\sqrt{10^{-5}/1} = 0.0032$.

Several important points summarize mutation loads:

- mutational loads are found in essentially all species, including threatened species
- deleterious alleles are normally found only at low frequencies, typically much less than 1% at any locus
- deleterious alleles are found at many loci
- there are differences in frequencies of deleterious alleles according to the mode of inheritance and dominance
- most deleterious mutations are partially recessive

SUGGESTED FURTHER READING

Frankham, R., J. D. Ballou. & D. A. Briscoe. 2002. *Introduction to Conservation Genetics*. Cambridge University Press, Cambridge, UK. Chapters 6, 7 and 9 have extended treatments of these topics, plus references.

Briggs, D. & S. M. Walters. 1997. *Plant Variation and Evolution*, 3rd edn. Cambridge University Press, Cambridge, UK. Reviews evidence for adaptive genetic changes in plants.

Falconer, D. S. & T. F. C. Mackay. 1996. *Introduction to Quantitative Genetics*, 4th edn. Longman, Harlow, UK. This textbook provides a very clear treatment of the topics in this chapter with a focus on animal breeding applications.

Futuyma, D. J. 1998. *Evolutionary Biology*, 3rd edn. Sinauer, Sunderland, MA. A textbook with a broad readable coverage of evolution, adaptations and the genetic processes underlying them.

Hartl, D. L. & A. G. Clarke. 1997. *Principles of Population Genetics*, 3rd edn. Sinauer, Sunderland, MA. A widely used textbook on population genetics.

Hedrick, P. W. 2000. *Genetics of Populations*, 2nd edn. Jones & Bartlett, Boston, MA. This textbook has extensive, authoritative treatments of many of the topics in this chapter.

Chapter 4

Genetic consequences of small population size

Populations of conservation concern are small and/or declining in numbers. Small, isolated populations suffer accelerated inbreeding and loss of genetic diversity leading to reduced reproductive fitness (inbreeding depression) and reduced ability to evolve in response to environmental change

Terms

Bottleneck, effective population size, evolutionary potential, F statistics, fixation, harmonic mean, idealized population, identical by descent, inbreeding, inbreeding coefficient, inbreeding depression, metapopulation, Poisson distribution, random genetic drift, stochasticity

Mauritius kestrel: a species that survived a population size bottleneck of a single pair, but one with genetic 'scars'.

Importance of small populations in conservation biology

Small or declining populations are more prone to extinction than large stable populations. Species whose adult population sizes are stable and less than 50, 250 or 1000 are, respectively, designated as critically endangered, endangered and vulnerable (Chapter 1). Only ~100 (adults plus juveniles) critically endangered northern hairy-nosed wombats survive in Australia, while the Mauna Kea silversword in Hawaii declined to about two dozen plants. Some species have reached such low numbers that they exist, or have existed, only in captivity. These include Arabian oryx, black-footed ferret, European bison, Père David's deer, Przewalski's horse, scimitar-horned oryx, California condor, Guam rail, Cook's kok'io plant, Franklin tree, and Malheur wirelettuce.

Some species have experienced population size reductions (**bottlenecks**), but have since recovered. The Mauritius kestrel declined to a single pair but has now recovered to 400-500 birds (Box 4.1). Northern elephant seals were reduced to 20-30 individuals but now number over 150 000. These populations pay a genetic cost for their bottlenecks; they typically have higher levels of inbreeding, lower reproductive fitness, reduced genetic diversity and compromised ability to evolve (Box 4.1).

> Species of conservation concern have, by definition, small or declining population sizes

Guam rail

| Box 4.1 | A population size bottleneck in the Mauritius kestrel and its genetic consequences (after Groombridge et al. 2000) |

The decline of the Mauritius kestrel began with the destruction of native forest and its plunge towards extinction resulted from thinning of egg shells and greatly reduced hatchability following use of DDT insecticide, beginning in the 1940s. By 1974, its population numbered only four individuals, with the subsequent population descending from only a single breeding pair. Under intensive management the population grew to 400–500 birds by 1997.

While this is a success story, the restored Mauritius kestrel carries genetic scars from its near extinction. It now has a very low level of genetic diversity for microsatellite loci, with 72% lower allelic diversity and 85% lower heterozygosity than the mean for non-endangered kestrels (see below). Prior to its decline, the Mauritius kestrel had substantial genetic diversity, based on museum skins from 1829–1894, although this was lower than that of non-endangered kestrel species.

Species	A	H_e
Endangered		
Mauritius kestrel		
Restored	1.41	0.10
Museum skins	3.10	0.23
Non-endangered		
European kestrel	5.50	0.68

The reproductive fitness of the Mauritius kestrel has been adversely affected by inbreeding in the early post-bottleneck population, with lowered fertility and productivity than in comparable falcons, and higher adult mortality in captivity.

Loss of genetic diversity

Genetic diversity is lost when population sizes are reduced

Many endangered species have suffered bottlenecks (Table 4.1) or long periods at small population sizes. Genetic diversity is typically lost as a consequence, reducing evolutionary potential. For example, the northern elephant seals have no allozyme diversity. Such losses are a consequence of the random sampling, or chance, nature of transmission of alleles from one generation to the next.

Arabian oryx

Table 4.1 Bottlenecks in endangered species (numbers of founders breeding in captivity)

Species	Bottleneck size
Mammals	
Arabian oryx	10
Black-footed ferret	7
European bison	13
Indian rhinoceros	17
Père David's deer	~5
Przewalski's horse	12
Snow leopard	7
Birds	
California condor	14
Guam rail	12
Mauritius kestrel	2
Nene (Hawaiian goose)	17
Puerto Rican parrot	13
Whooping crane	14
Invertebrates	
British field cricket	12

Source: See Frankham *et al.* (2002).

Whooping crane

Chance effects arise from random sampling of gametes in small populations

Chance effects and genetic drift

When sexual diploid populations reproduce, the subsequent generation is derived from a sample of parental gametes. In small populations, some alleles, especially rare ones, may not be transmitted just by chance. The frequencies of alleles that are transmitted to the following generation are likely to differ from those in the parents (Fig. 4.1). Over multiple generations allele frequencies fluctuate from one generation to the next, a process termed **random genetic drift**.

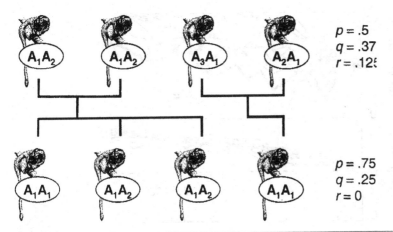

$p = .5$
$q = .37$
$r = .12\!5$

$p = .75$
$q = .25$
$r = 0$

Fig. 4.1 Genetic drift in allele frequencies in a small population of golden lion tamarins. p, q and r are the frequencies of alleles A_1, A_2 and A_3, respectively. Allele A_3 is lost by chance. Further, the frequencies of A_1 and A_2 change from one generation to the next, with A_1 rising and A_2 falling.

Genetic drift

It may seem that chance effects would have only minor impacts on the genetic composition of populations. However, random sampling of gametes within small populations has three consequences of major significance in evolution and conservation:

> Genetic drift has major impacts on the evolution of small populations

- loss of genetic diversity and fixation of alleles within populations, with consequent reduction in evolutionary potential
- diversification among replicate populations from the same original source (e.g. fragmented populations)
- genetic drift overpowering natural selection.

These features are exemplified in the flour beetle populations in Fig. 4.2. First, individual populations show genetic drift. For example, in the $N = 10$ population marked with an asterisk, the frequency of the allele begins at 0.5, drops for three generations, and then rises and falls until generation 20 when its value is approximately 0.65. Note that the fluctuations in frequencies for populations of size $N = 10$ are much greater than for populations with $N = 100$, clearly illustrating that genetic drift is greater in smaller populations.

> The effects of chance are greater in small than in larger populations

Second, some populations lose genetic diversity and reach **fixation** (homozygosity). Seven of the 12 $N = 10$ populations became homozygous over 20 generations. Six of the seven populations became fixed for the wild-type allele and one for the black allele. None of the $N = 100$ populations became fixed over the 20 generations.

Third, there is random diversification among replicate populations, particularly in the $N = 10$ populations. These populations all began with frequencies of 0.5, yet end up with frequencies ranging from 0 to 1. Again, the diversification among replicate populations is much less for the $N = 100$ populations.

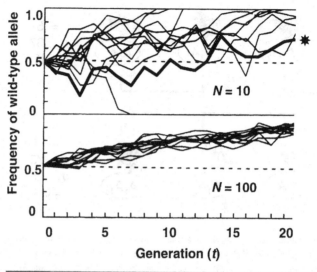

Fig. 4.2 Random genetic drift of the wild-type (+) and black (*b*) alleles at a body colour locus in the red flour beetle. Two population sizes, *N* = 10 and *N* = 100, were used, with 12 replicates of each. All populations began with frequencies of 0.5 for the two alleles and were maintained by random sampling of either 10, or 100 individuals to be parents of each succeeding generation (after Rich *et al.* 1979). *Large variation in allele frequencies occurred in the small (N = 10) populations due to random genetic drift, both among replicates and from generation to generation in individual replicates. Conversely, allele frequencies in the large populations showed greater consistency.*

Finally, it is clear from the overall pattern that natural selection favoured the wild-type allele (+) over the mutant black (*b*). Nonetheless, genetic drift has overpowered this selection and resulted in fixation of *b* in one *N* = 10 population.

Natural, fragmented populations will experience these effects at all their genetic loci, with smaller fragments experiencing greater genetic drift than larger fragments.

Population bottlenecks result in loss of alleles (especially rare ones), reduced genetic diversity and random changes in allele frequencies

The impact on allele frequencies of reducing population size to a single pair of parents in experimental populations of fruit flies is shown in Fig. 4.3. Note the loss of alleles, particularly of rare alleles, and that allele frequencies have changed from those in the parent population. Replicate bottlenecked populations varied in the alleles they lost and in the frequencies of the alleles they retained. On average, heterozygosity dropped from 0.61 in the base population to 0.44 in the bottlenecked populations, and the number of alleles from 12 to 3.75. Note that the cumulative effect of maintaining populations at a size of 100 for 57 generations has resulted in a similar loss of genetic diversity, an issue we address below.

Loss of heterozygosity following bottlenecks

The proportion of initial heterozygosity retained after a single generation bottleneck is

$$\frac{H_1}{H_0} = 1 - \frac{1}{2N} \tag{4.1}$$

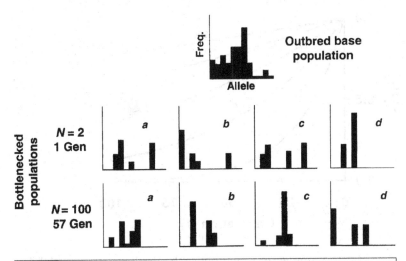

Fig. 4.3 Effect of population bottlenecks on experimental populations of fruit flies. The distribution of allele frequencies at a microsatellite locus in the large outbred base population, in four replicate populations subjected to a bottleneck of one pair of flies, and in four populations maintained at a size of 100 for 57 generations (England 1997). *Alleles are lost, especially rare ones, and allele frequencies distorted in the bottlenecked populations.*

where H_1 is the heterozygosity immediately after the bottleneck, and H_0 that before. A proportion $1/(2N)$ of the original heterozygosity is lost. Thus, single generation bottlenecks have to be severe before they have a substantial impact on heterozygosity. A bottleneck of size 2 still retains 75% of initial heterozygosity. Heterozygosity declined from 0.61 to 0.44 following the bottleneck in the experiment described in Fig. 4.3, similar to this prediction.

A bottleneck of $N = 25$ only reduces heterozygosity by 2%, while a bottleneck of 100 reduces it by only 0.5%. Loss of genetic diversity arises predominantly from sustained reductions in population size, rather than single generation bottlenecks. In the Mauritius kestrel, heterozygosity declined 57% from 0.23 to 0.10 as a result of a single pair bottleneck (Box 4.1). While this was greater than expected, additional genetic diversity would have been lost during the six generations the population spent at sizes of less than 50. Many threatened wildlife populations show evidence of loss of genetic diversity due to population size bottlenecks of various durations (Table 2.7).

Effects of sustained population size restrictions on genetic diversity

The impact of reduced population size occurs in every generation and losses accumulate with time. If population size is constant in each generation, equation 4.1 can be extended to obtain an expression for the effects of sustained population size restriction on heterozygosity, as follows:

Loss of heterozygosity is a process of continuous decay that is more rapid in smaller than larger populations

$$\frac{H_t}{H_0} = \left(1 - \frac{1}{2N}\right)^t \qquad (4.2)$$

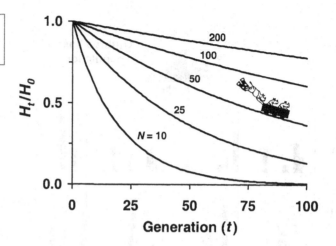

Fig. 4.4 Predicted decline in heterozygosity over time in different-sized populations.

where t is the number of generations. The predicted exponential declines in heterozygosity with time (measured in generations) in different sized populations are shown in Fig. 4.4. All populations are expected to lose genetic diversity, but the rate of loss is greater in smaller than in larger populations.

Example 4.1 demonstrates that populations of size 500 will lose only about 5% of their initial heterozygosity over 50 generations, while populations with $N = 25$ lose 64% of their initial heterozygosity. Further, the shorter the generation length, the more rapid in absolute time will be the loss. Consequently, similar-sized populations of black-footed ferrets with generation lengths of two years will lose genetic diversity more rapidly than elephants with generation lengths of 26 years.

Example 4.1	Expected loss of heterozygosity due to sustained population size reduction

From equation 4.2, the expected proportion of heterozygosity retained over 50 generations in a population of size 500 is:

$$\frac{H_t}{H_0} = \left(1 - \frac{1}{2N}\right)^t = \left(1 - \frac{1}{(2 \times 500)}\right)^{50} = \left(\frac{999}{1000}\right)^{50} = 0.951$$

For a population with $N = 25$, the proportion of initial heterozygosity retained at generation 50 is expected to be

$$\frac{H_t}{H_0} = \left(1 - \frac{1}{2N}\right)^t = \left(1 - \frac{1}{(2 \times 25)}\right)^{50} = \left(\frac{49}{50}\right)^{50} = 0.364$$

Inbreeding

Inbreeding is unavoidable in small populations and leads to reductions in reproduction and survival

So far we have considered the effects of small population size on loss of genetic diversity and consequent loss of potential to evolve. Reduced population size also has acute effects resulting from inbreeding.

In small closed populations, matings among relatives (inbreeding) is inevitable. With time, every individual becomes related so that matings between unrelated individuals are impossible. This is not a result of active choice of matings among relatives, but simply a consequence of the small number of founders and the small population size. Inbreeding also becomes inevitable in larger populations, but it takes longer.

Inbreeding is of profound importance in conservation biology as it leads to reductions in heterozygosity, to reduced reproduction and survival (reproductive fitness), and to increased risk of extinction (Chapter 5). Loss of reproductive fitness as a consequence of inbreeding is referred to as **inbreeding depression**. For example, in a study of 44 captive mammal populations, inbred individuals experienced higher juvenile mortality than outbred individuals in 41 cases. On average, brother–sister mating resulted in a 33% reduction in juvenile survival.

There is now overwhelming evidence that inbreeding also adversely affects wild populations. For example, in a meta-analysis of 157 data sets from 34 taxa, inbred individuals had poorer attributes than comparable outbreds in 141 cases (90%), two were equal and only 14 were in the opposite direction. Inbreeding is deleterious across a diversity of animals and plants, including golden lion tamarins, lions, greater prairie chickens, Mexican jays, song sparrows, American kestrels, reed warblers, Atlantic salmon, desert topminnows, rainbow trout, and many species of plants.

Measuring inbreeding: inbreeding coefficient (F)

The primary consequence of matings between relatives is that their offspring have an increased probability of inheriting alleles that are recent copies of the same DNA sequence (**identical by descent**). For example, in Fig. 4.5 the A_1A_1 or A_2A_2 offspring resulting from self-fertilization have inherited two alleles which are identical copies of the A_1 or A_2 alleles in their parent. The two identical copies of an allele do not need to come from an individual in the previous generation, but may come from an ancestor in a more remote generation. In Fig. 4.5 an offspring resulting from a brother–sister mating may inherit two copies of allele A_1, A_2, A_3 or A_4 from its grandparent. The grandparents are said to be **common ancestors**, as they are ancestors of both the mother and the father of the individual.

The probability that alleles at a locus uniting in an individual are identical by descent is termed the **inbreeding coefficient** (F). As F is a probability, it ranges from 0 in outbreds to 1 in completely inbred individuals.

To calculate inbreeding coefficients from first principles, each non-inbred ancestor is labelled as having unique alleles (A_1A_2, A_3A_4, etc.) (Fig. 4.5). The probability that an individual inherits two alleles identical by descent (A_1A_1, A_2A_2, etc) is computed from the paths of inheritance, assuming normal Mendelian segregation. For example, with

> The inbreeding coefficient of an individual (F) is the probability that it carries alleles at a locus that are identical by descent

Fig. 4.5 Inbreeding coefficients for individuals resulting from self-fertilization and full-sib mating.

Self-fertilization

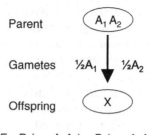

Parent A₁ A₂

Gametes ½A₁ | ½A₂

Offspring X

$$F = Pr(x = A_1A_1) + Pr(x = A_2A_2)$$
$$= ¼ + ¼ = ½$$

Full-sib mating

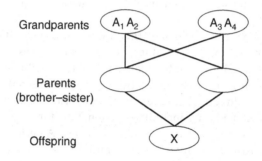

Grandparents A₁ A₂   A₃ A₄

Parents (brother–sister)

Offspring X

$$F = Pr(x = A_1A_1 \text{ or } A_2A_2 \text{ or } A_3A_3 \text{ or } A_4A_4)$$
$$= 1/16 + 1/16 + 1/16 + 1/16 = ¼$$

selfing the inbreeding coefficient is the probability that offspring X inherits either two A_1 alleles, or two A_2 alleles. Individual X has $½$ chance of inheriting A_1 in the ovule and $½$ chance of inheriting it through the pollen. Consequently, the probability that X inherits two identical A_1 alleles is $½ \times ½ = ¼$. The chance that X inherits two A_2 alleles is also $½ \times ½ = ¼$. The inbreeding coefficient is then the probability of inheriting either A_1A_1 or $A_2A_2 = ¼ + ¼ = 0.5$. Similarly, an offspring from a brother–sister (full-sib) mating has $F = 0.25$.

Genetic consequences of inbreeding

Inbreeding increases levels of homozygosity and exposes deleterious recessive alleles

Inbreeding may also refer to a mating system where related individuals mate at a rate greater than expected by random mating, such as in self-fertilization, brother–sister mating, etc. When this happens, the frequency of heterozygotes is reduced, and that of homozygotes increased, relative to Hardy–Weinberg expectations. We can see how this arises in the most extreme form of inbreeding, self-fertilization, by following the genotype frequencies expected under Mendelian inheritance (Fig. 4.6). If an A_1A_2 individual self-fertilizes, heterozygosity is halved in its progeny, and continues to halve in each subsequent generation.

Genotype frequencies (%)

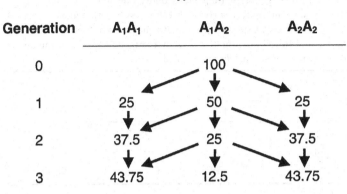

Generation	A_1A_1	A_1A_2	A_2A_2
0		100	
1	25	50	25
2	37.5	25	37.5
3	43.75	12.5	43.75

Fig. 4.6 Effect of self-fertilization on genotype frequencies. *The frequency of heterozygotes halves with each generation of selfing.*

Consequently, deficiencies of heterozygotes in populations, compared to Hardy–Weinberg equilibrium expectations, indicate non-random mating.

This reduction in heterozygosity due to inbreeding is directly related to the inbreeding coefficient F. Table 4.2 shows the effects of inbreeding, at a level of F, on genotype frequencies. When there is no inbreeding (F = 0), genotype frequencies are in Hardy–Weinberg equilibrium. When F = 0.5, the proportion of heterozygotes in the population is reduced by 50%.

Inbreeding does not directly alter allele frequencies. Rather it alters genotype frequencies. Inbreeding is often associated with small populations and allele frequencies change in such populations due to genetic drift, as discussed earlier.

A major practical consequence of inbreeding is that homozygotes for deleterious recessives become more frequent than in a random mating population. This is the primary cause of inbreeding depression. Since naturally outbreeding populations contain deleterious alleles (mostly partially recessive) at low frequencies in mutation-selection balance, inbreeding increases the risks of exposing them as homozygotes, as illustrated for chondrodystrophy in California condors in Table 4.3. The frequency of homozygous *dw/dw* is more than doubled in a population with an inbreeding coefficient of 25%. This effect occurs at all loci with deleterious alleles and can have a very large cumulative effect on the fitness and health of individuals.

Inbreeding reduces the frequency of heterozygotes in proportion to F, and exposes rare deleterious alleles

Table 4.2 Genotype frequencies under random mating compared to those in populations with inbreeding coefficients of F.

			Genotypes	
Population	F	+/+	+/m	m/m
Random mating	0	p^2	$2pq$	q^2
Partially inbred	F	$p^2 + Fpq$	$2pq\,(1 - F)$	$q^2 + Fpq$

Table 4.3 Expected genotype frequencies under inbreeding at the chondrodystrophy locus in California condors. The recessive lethal allele has a frequency of about 0.17. Expected genotype frequencies are shown for random mating and full-sib mating (F = 0.25) determined using the formulae in Table 4.2. The homozygous *dw/dw* chicks die

	Genotypes		
	+/+	*+/dw*	*dw/dw*
Random mating	0.6889	0.2822	0.0289
Partially inbred: F = 0.25	0.7242	0.2116	0.0642

F values can be directly calculated by comparing observed (H_o) to expected (H_e) heterozygosities in a population. As $H_o = (1 - F)\, 2pq$ and $H_e = 2pq$ (Table 4.2), then:

$$F = 1 - \frac{H_0}{H_e} \tag{4.3}$$

This relationship can also be used to measure the accumulated affect of inbreeding over time if heterozygosity at generation t (H_t) is compared to initial heterozygosity (H_0):

$$F = 1 - \frac{H_t}{H_0} \tag{4.4}$$

Pedigrees

Where available, pedigrees can be used to determine the inbreeding coefficient of an individual. This allows us to evaluate the effects of inbreeding on survival or reproduction rates, etc. using data from individuals with different levels of inbreeding. Computation of F from first principles becomes impractical when dealing with complex pedigrees. Consequently, simpler alternative methods have been devised for these situations.

The inbreeding coefficient can be obtained by counting the individuals in the path from one parent, through the common ancestor to the other parent (including both parents), and raising $\frac{1}{2}$ to this power. The $\frac{1}{2}$ reflects the probability that an allele is transmitted one generation to the next, and the power reflects the number of such parent–offspring steps in the pedigree. For example, in Fig. 4.7 there are three individuals connecting parents of individual X through their common ancestor A, i.e. through D, A and E. The inbreeding coefficient of X is $(\frac{1}{2})^3$, if A is not itself inbred.

In more complex pedigrees, the parents of an individual may be related through more than one common ancestor, or from the same ancestor through different paths. Each common ancestor, and each path, contributes an additional probability of the progeny having

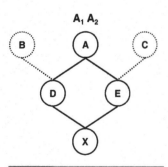

Fig. 4.7 Pedigree with mating between half-sibs.

identity by descent. The inbreeding coefficient is the sum of all the probabilities, as follows

$$F = \Sigma(\tfrac{1}{2})^n(1 + F_{ca}) \tag{4.5}$$

where n is the number of individuals in the path from one parent to a common ancestor and back to the other parent, and F_{ca} is the inbreeding coefficient of that particular common ancestor. These contributions to inbreeding are summed for each different path linking both parents to each common ancestor.

We apply this method to the simple pedigree in Fig. 4.8. The individuals to count from one parent to the other through the common ancestor (A) are F, D, B, A, C, E and G, making $n = 7$ steps. Thus F_X in Fig. 4.8 is $(\tfrac{1}{2})^7 (1 + F_A)$, being 1/128 if individual A is not inbred.

For complex pedigrees, inbreeding coefficients are typically calculated using computer programs.

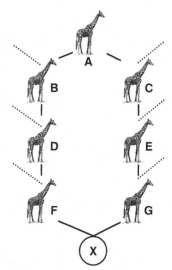

Fig. 4.8 Pedigree with a more remote common ancestor. The dotted lines represent paths to other ancestors that are not on the path to the common ancestor A.

Inbreeding in small random mating populations

While a minority of plants routinely self-fertilize, animals normally do not self. Despite many opportunities for relatives to mate due to the physical proximity of siblings, offspring and parents, inbred matings are generally rarer than expected from proximity. Many species have evolved inbreeding-avoidance mechanisms. Indeed two of the most variable genetic loci known, the self-incompatibility (SI) loci in plants and the major histocompatibility (MHC) loci in vertebrates, are associated with inbreeding avoidance.

In naturally outbreeding species, inbreeding arises predominantly from small population size

In a very large random mating population, inbreeding is close to zero as there is very little chance of mating with a relative. However, in small closed populations every individual eventually becomes related by descent and inbreeding is unavoidable. By combining the concepts in Equations 4.2 and 4.4 we can derive an expression for inbreeding in a closed random mating population after t generations, namely:

$$F_t = 1 - \frac{H_t}{H_0} = \left(1 - \frac{1}{2N}\right)^t \tag{4.6}$$

Thus, inbreeding accumulates with generations in all closed finite populations, at a rate dependent on their sizes (Fig. 4.9).

Example 4.2 illustrates the rapid accumulation of inbreeding in a small closed population with only four individuals per generation. The population reaches an average inbreeding coefficient of 74% by generation 10. This is approximately equivalent to the level of inbreeding due to two generations of selfing or six generations of full-sib mating – but it was achieved in a random mating population. Since captive populations of endangered species within individual zoos are often of this size, individuals have to be moved between institutions if rapid inbreeding is to be minimized (Chapter 8).

Inbreeding accumulates over time and is more rapid in smaller than in larger populations

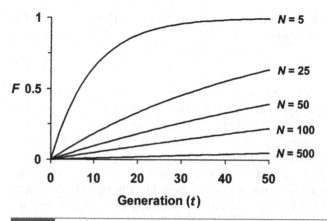

Fig. 4.9 Increase in inbreeding coefficient F with time in finite populations of different sizes (N). *Inbreeding increases more rapidly in smaller than in larger populations.*

Example 4.2 | Accumulation of inbreeding in a small closed captive population

Many captive populations of threatened species in individual zoos are small, and would accumulate inbreeding rapidly if they were kept closed, i.e. no individuals exchanged between zoos. If a zoo started a breeding program with four unrelated individuals, and kept the breeding population at four parents per generation over many generations, the inbreeding coefficient would increase as follows:

Generation 0 $\quad F = 0$

Generation 1 $\quad F = 1 - \left(1 - \dfrac{1}{2N}\right)^1 = 1 - \left(1 - \dfrac{1}{8}\right) = 0.125$

Generation 2 $\quad F = 1 - \left(1 - \dfrac{1}{2N}\right)^2 = 1 - \left(1 - \dfrac{1}{8}\right)^2 = 0.23$

Generation 3 $\quad F = 1 - \left(1 - \dfrac{1}{2N}\right)^3 = 1 - \left(1 - \dfrac{1}{8}\right)^3 = 0.33$

Generation 10 $\quad F = 1 - \left(1 - \dfrac{1}{2N}\right)^{10} = 1 - \left(1 - \dfrac{1}{8}\right)^{10} = 0.74$

Thus, the inbreeding coefficient increases rapidly and reaches 74% by generation 10.

Measuring population size

Until this time we have discussed the impacts of size on genetic processes within populations as though these are related to the absolute, or census size. This is rarely the case. Many populations of small mammals and insects fluctuate wildly in size, and it is the minimum size that most profoundly affects genetic processes (see later). Similarly, most populations contain juveniles and post-reproductives. Amongst reproductives, there may be considerable variation in their contributions to subsequent generations. Further, species vary in mating system (e.g. monogamy, harems), and from approximately random mating to selfing and asexual reproduction.

The same number of individuals may result in very different genetic impacts in different species, depending on population structure and breeding system. Consequently, we must define precisely what we mean by population size in conservation genetics. We do this by comparing real populations to a hypothetical **idealized population**.

The idealized population

We begin by assuming a large (essentially infinite) random mating base population, from which we take a sample of size N adults to form the ideal population (Fig. 4.10). This population is maintained as a random mating, closed population in succeeding generations. Alleles may be lost by chance, and allele frequencies may fluctuate due to genetic drift. The simplifying conditions applying to the idealized population are:

> The idealized population, to which all other populations are compared, is a closed random mating population of hermaphrodites with discrete generations, constant size, and Poisson variation in family sizes

- the number of breeding individuals is constant in all generations
- generations are distinct and do not overlap
- there is no migration or gene flow
- all individuals are potential breeders
- all individuals are hermaphrodites
- union of gametes is random, including the possibility of selfing
- there is no selection at any stage of the life cycle
- mutation is ignored
- the number of offspring per adult averages 1, and has a variance of 1.

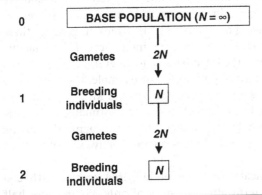

Fig. 4.10 Idealized population. From a very large base population a hypothetical sample of N adults is taken and this population is maintained as a random mating, closed population with constant numbers of parents in each generation.

Within the population, breeding individuals contribute gametes equally to a pool from which zygotes are formed. Survival of zygotes is random, so that the contributions of adults to the next generation are not equal. The mean number of offspring per adult is 1, but may vary from 0, 1, 2, 3, 4, etc. for different individuals, according to the terms of the **Poisson distribution**.

The characteristics of the idealized population are well defined because of these assumptions and a large body of theory has been derived for them. Consequently, by equating real populations to the idealized population, theory can be utilized to make practical predictions.

Effective population size (N_e)

We can standardize the definition of population size by describing it in terms of its **effective population size** (N_e). The effective size of a population is the size of an idealized population that would lose genetic diversity (or become inbred) at the same rate as the actual population. For example, if a real population loses genetic diversity at the same rate as an ideal population of 100, then we say it has an effective size of 100, even if it has a census size of 1000 individuals. Thus, the N_e of a population is a measure of its genetic behaviour, relative to that of an ideal population.

> Genetic processes in small populations depend on the effective population size rather than on the number of individuals

All of the adverse genetic outcomes of small population size depend on the effective population size, rather than on the absolute number of individuals. Consequently, N_e should be used in place of N in equations 4.1, 4.2 and 4.6. As we will see, in practice, the effective size of a population is usually significantly less than the number of breeding adults.

If a real population had all of the properties of an ideal population then $N_e = N$. However, any characteristics of a real population that deviate from those of an ideal population will cause the census size to be different from N_e. Real populations deviate in structure from the assumptions of the idealized population by having unequal sex ratios, high variation in family sizes, variable numbers in successive generations, and in having overlapping generations. In general, these factors deviate such that $N_e < N$. The combined effect of these factors has been found to reduce the effective size to an average of 11% of the census size in unmanaged populations. For example, the threatened winter run of chinook salmon in the Sacramento River of California has about 2000 adults, but its effective size was estimated to be only 85 ($N_e/N = 0.04$), much lower than previously recognized. Genetic concerns are much more immediate with an effective size of 85 than with 2000.

> Estimates of effective population size that encompass all relevant factors average only 11% of census sizes

The sobering implication is that endangered species with 250 adults may have long-term effective sizes of only 25, and lose half of their current heterozygosity in 34 generations. By this time, the population will have become inbred to the point where inbreeding depression increases extinction risks (Chapter 5).

Measuring effective population size

Since $N_e \ll N$ we need to estimate the impacts on effective population size of unequal sex-ratio, variation in family size, and fluctuations in population size over generations. Unequal sex-ratios have least effect, and fluctuations in population size the greatest effect. No clear or consistent differences have been found among major taxonomic groups.

> Effective population size can be estimated from demographic data on sex-ratio, variance in family sizes and fluctuations in population size over generations, or from genetic data

Unequal sex-ratio

In many wild populations the number of breeding females and males are not equal. Many mammals have harems where one male mates with many females, with many males making little or no genetic contribution to the next generation. This occurs in an extreme form in elephant seals where a single male may maintain a hundred or more females in his harem. In a few species, the situation is reversed. The equation accounting for the effects of unequal sex-ratio is

> Unequal sex-ratios reduce the effective size of the population towards the number of the sex with fewer breeding individuals

$$N_e \sim \frac{4\,N_{ef}\,N_{em}}{(N_{ef} + N_{em})} \tag{4.7}$$

where N_{ef} is the effective number of breeding females and N_{em} the effective number of breeding males. This is the single-generation effective population size due to sex-ratio alone, all other characteristics are assumed to conform to an idealized population.

As sex-ratios deviate from 1:1, the N_e/N ratio declines. For example, an elephant seal harem with one male and 100 females has an effective size of only 4 (Example 4.3).

Example 4.3	Reduction in effective size due to unequal sex-ratio in elephant seals

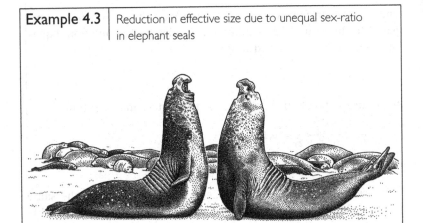

If a harem has one male and 100 females, the effective size is

$$N_e \sim \frac{4\,N_{ef}N_{em}}{(N_{ef} + N_{em})} = \frac{4 \times 100 \times 1}{100 + 1} = 3.96$$

Thus the effective size of the harem is 3.96, approximately 4% of the actual size of 101.

Variation in family size

Family sizes (lifetime production of offspring per individual) in wild populations typically show greater variation than that expected (Poisson) for the idealized population. In a stable population of a randomly breeding monogamous species, the mean family size (k) is 2 (an average of 1 male and 1 female to replace both parents) and the variance (V_k) is 2. Note that the variance equals the mean for a Poisson distribution. Most species in the wild have V_k/k ratios well in excess of the value of 1 assumed for the idealized population. High variation in family sizes in wildlife is partly due to individuals that contribute no offspring to the next generation. Similar but less extreme effects arise from very large and very small families.

The effect of variation in family sizes in a population otherwise having the structure of an idealized population is

$$N_e \sim \frac{4N}{(V_k + 2)} \tag{4.8}$$

This is the single-generation effective population size due to family size alone.

This equation indicates that the higher the variance in family size, the lower the effective population size. For example, in Darwin's cactus finch high variance in family size (6.74), compared to the Poisson expectation (2) reduces the effective population size to less than half the number of breeding pairs (Example 4.4).

Example 4.4 | Reduction in effective population size through high variance in family size in Darwin's large cactus finch

The variance in family sizes for Darwin's cactus finch is 6.74, compared to the value of 2 assumed for an idealized population. Equation 4.8 can be rearranged to give

$$\frac{N_e}{N} \sim \frac{4}{(V_k + 2)}$$

If we insert the observed value into this equation, we obtain

$$\frac{N_e}{N} \sim \frac{4}{(6.74 + 2)} = 0.46$$

If breeding animals are managed so that all produce the same number of offspring (i.e. family sizes are equalized), $V_k = 0$. In a stable population of a monogamous species this corresponds to each family contributing 1 male and 1 female as parents of the next generation. By substitution $V_k = 0$ into equation 4.7 we obtain $N_e \sim 2N$. Thus, the effective size of a population can be approximately double the number of parents. When all families contribute alleles equally to the next generation, there is minimal distortion in allele frequencies and the proportion of the genetic diversity passed on is maximized. Further, there is a better choice of mates and inbreeding is minimized.

This observation is of critical importance to management of captive populations. Equalization of family sizes potentially allows the limited captive breeding space for endangered species to be effectively doubled (Chapter 8).

Fluctuations in population size

The greatest reductions in N_e/N ratios arise from fluctuations in population size over generations. Wild populations vary in numbers as a consequence of variation in food availability, climatic conditions, disease epidemics, catastrophes, predation, etc. For example, lynx and snowshoe hare populations fluctuate in size, the hare showing about a 30-fold difference between high and low years and the lynx about an 80-fold difference. Further, population declines of 70–90% due to droughts, diseases, etc. are not uncommon for large mammals.

Fluctuations in population size, over generations, reduce N_e below the average number of adults

The effective size of a fluctuating population is not the arithmetic average, but the **harmonic mean** of the effective population sizes over t generations:

$$N_e \sim \frac{t}{\Sigma(1/N_{ei})} \qquad (4.9)$$

Lynx

where N_{ei} is the effective size in the ith generation and N_e is the long-term, overall effective population size.

The long-term effective size is closest to the smallest single generation N_e. For example, the northern elephant seal was reduced to 20–30 individuals, but has since recovered to over 150 000. Its effective population size over this time is about 60 (Example 4.5). This is far closer to the minimum population size than to the arithmetic average or the maximum. This relationship is best explained by noting that an allele lost in a generation of low population size is not regained when the population recovers. Similarly, inbreeding due to small population size is not reduced when the population increases in size.

Example 4.5 | Reduction in N_e due to fluctuations in population size

The northern elephant seal was reduced to 20–30 individuals by hunting. It has since recovered to over 150 000. For simplicity we assume that the effective population size declined from 150 000 to 20 and recovered to 150 000 in a 3-generation period. The effective size of the population is

$$N_e = \frac{t}{(1/N_{e1}) + (1/N_{e2}) + (1/N_{e3})}$$

$$= \frac{3}{(1/150\,000) + (1/20) + (1/150\,000)} = 60$$

The effective size of 60 is much closer to the minimum size than to the average size (100 006).

Effective population sizes can also be estimated using equations, such as 4.2, relating changes in genetic diversity with N_e.

Population fragmentation

The genetic consequences of population fragmentation depend critically upon the level of gene flow. With restricted gene flow, fragmentation is usually highly deleterious in the long term

Many populations throughout the world have become fragmented, frequently as a result of human actions. The impacts of population fragmentation on genetic diversity, differentiation, inbreeding and extinction risk depend on the level of gene flow among fragments. These in turn depend on the:

• number of population fragments
• distribution of population sizes in the fragments
• geographic distribution of populations
• distance among fragments
• dispersal ability of the species
• environment of the matrix among the fragments and its impact on dispersal
• time since fragmentation
• extinction and recolonization rates across fragments.

All of the issues of loss of genetic diversity and inbreeding depression, with respect to reduced population size, come into play when populations are fragmented. In fragmented populations with diminished gene flow, these adverse effects are usually more severe than in a non-fragmented population of the same total size.

The effects of population fragmentation are illustrated in Fig. 4.11, where four small isolated fragments (SS) are compared to a single large population (SL) of the same initial total size. The large population (1) possesses four alleles, A_1, A_2, A_3 and A_4. In the short term, the four SS populations (2) rapidly become homozygous and lose fitness through inbreeding depression. Loss of genetic diversity is slower in the single large population (3) – it only loses allele A_4. However, fixation in the SS populations is at random, so that overall all four alleles are retained, while the SL population has lost the A_4 allele. Thus, as long as there are no extinctions of SS populations, they retain greater overall allelic diversity than the SL population.

However, in the long term, extinction rates will be greater in smaller than in larger population fragments due to environmental and demographic stochasticity, catastrophes and genetic factors (Chapter 5). With extinction of some SS populations (4), the SL population retains more genetic diversity and has higher reproductive fitness than all the SS populations combined (now only two populations).

The endangered red-cockaded woodpecker in the southeastern USA illustrates many of the features and genetic problems associated with habitat fragmentation for a species with a once continuous distribution (Box 4.2). Isolated and small woodpecker populations show loss of genetic diversity compared to larger populations. Differentiation among populations in different patches is evident, with adjacent populations generally being more similar. The small populations are expected to suffer from inbreeding depression.

Initial populations

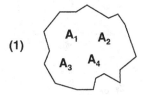

(1)

Short-term, no extinctions

Several small   **Single large**

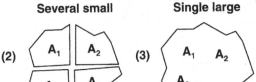

(2)   (3)

**Long-term, extinction of some
small populations**

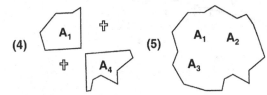

(4)   (5)

Fig. 4.11 The genetic consequences of a single large population (SL) versus several small (SS) completely isolated population fragments of initially the same total size over different time frames. (1) A_1–A_4 represent four alleles initially present in the population. In the short-term, without extinctions, the several small populations (2) are expected to go to fixation more rapidly, but to retain greater overall genetic diversity than the single large population (3). The chances are greater that an allele will be totally lost from the large population, than from all small populations combined. However, the SS populations will each be more inbred than the SL population. In the longer term, when extinctions of small, but not large, populations occur, the sum of the small surviving populations (4) will retain less genetic diversity than the single large population (5).

| **Box 4.2** | Impact of habitat fragmentation on the endangered red-cockaded woodpecker metapopulation in southeastern USA (Stangel et al. 1992; Kulhavy et al. 1995; Daniels et al. 2000) |

The red-cockaded woodpecker was once common in the mature pine forests of the southeast United States. It declined in numbers, primarily due to habitat loss, and was placed on the US endangered species list in 1970. It now survives in scattered and isolated sites within the US southeast (see map). A species recovery plan is being implemented to manage the species.

Red-cockaded woodpecker

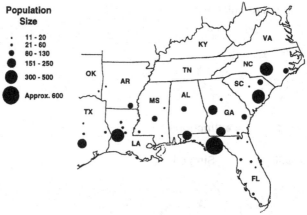

There is little migration among isolated sites. As predicted, smaller populations show the greatest loss of genetic diversity and the most divergence.

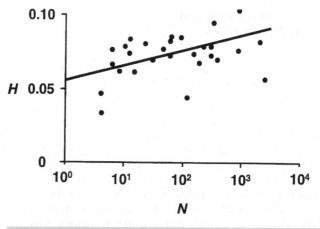

Moderate divergences in allele frequencies exist among woodpecker populations. There is a general tendency for closer genetic similarity among geographically proximate populations – shown in the cluster analysis of genetic distances (see Chapter 6) among 14 populations below.

Computer simulations (Chapter 5) indicate that the smallest woodpecker populations are likely to suffer from inbreeding depression in the near future.

In response to the threats posed by fragmentation, management of the woodpeckers involves habitat protection, improvement of habitat suitability by constructing artificial nest holes, reintroductions into suitable habitat where populations become extinct, and augmentation of small populations to minimize inbreeding and loss of genetic diversity. This is one of the most extensive management programs for a fragmented population anywhere in the world.

The degree of differentiation among fragments can be described by partitioning the overall inbreeding into components within and among populations (*F* statistics)

Measuring population fragmentation: *F* statistics

Inbreeding resulting from population fragmentation can be used to measure the degree of differentiation that has occurred among fragments. Differentiation among fragments or sub-populations is directly related to the inbreeding coefficients within and among populations. The inbreeding in the total population (F_{IT}) can be partitioned into that due to:

- inbreeding of individuals relative to their sub-population or fragment, F_{IS}, and
- inbreeding due to differentiation among sub-populations, relative to the total population, F_{ST}.

F_{IT}, F_{IS} and F_{ST} are referred to as *F* **statistics**. F_{IS} is the inbreeding coefficient, *F*, averaged across all individuals from all population fragments. F_{ST} is the effect of the population subdivision on inbreeding. With high rates of gene flow among fragments, F_{ST} is low. With low rates of gene flow among fragments, populations diverge and become inbred, and F_{ST} increases.

The F statistics can be calculated using equation 4.3 which relates heterozygosity and inbreeding. This allows F statistics to be determined from heterozygosity for genetic markers using the following equations:

$$F_{IS} = 1 - \left(\frac{H_I}{H_S}\right) \tag{4.10}$$

$$F_{ST} = 1 - \left(\frac{H_S}{H_T}\right) \tag{4.11}$$

$$F_{IT} = 1 - \left(\frac{H_I}{H_T}\right) \tag{4.12}$$

where H_I is the observed heterozygosity averaged across all population fragments, H_S is the expected Hardy–Weinberg heterozygosity averaged across all population fragments, and H_T is the expected Hardy–Weinberg heterozygosity for the total population (equivalent to H_e). F_{ST} ranges from 0 (no differentiation between fragments) to 1 (fixation of different alleles in fragments).

Example 4.6 illustrates the calculation of the F statistics based on heterozygosities for the endangered Pacific yew in western North America. This species exhibits inbreeding within populations ($F_{IS} > 0$) and differentiation among populations ($F_{ST} > 0$).

Example 4.6	Computation of F statistics for the rare Pacific yew (data from El-Kassaby & Yanchuk 1994)

This example is based on genotype frequencies and heterozygosities for 21 allozyme loci in nine Canadian populations. Average observed heterozygosity (H_I) across the nine populations was 0.085, while the average expected heterozygosity for these populations (H_S) was 0.166. Consequently, inbreeding within populations F_{IS} is

$$F_{IS} = 1 - \left(\frac{H_I}{H_S}\right) = 1 - \left(\frac{0.085}{0.166}\right) = 0.49$$

This is a high level of inbreeding, but it is not due to selfing as the species has separate sexes. It is probably due to clustering of relatives (offspring growing close to their parents and clumping of relatives from bird and rodent seed caches).

Pacific yew

The expected heterozygosity averaged across all nine populations (H_T) was 0.18, so inbreeding due to population differentiation (F_{ST}) is

$$F_{ST} = 1 - \left(\frac{H_S}{H_T} \right) = 1 - \left(\frac{0.166}{0.180} \right) = 0.078$$

This indicates only a modest degree of population differentiation. The total inbreeding due to both inbreeding within populations and differentiation among them (F_{IT}) is

$$F_{IT} = 1 - \left(\frac{H_I}{H_T} \right) = 1 - \left(\frac{0.085}{0.18} \right) = 0.53$$

F_{ST} values are generally inversely related to dispersal ability. Based on allozyme data, they average 0.25–0.3 for mammals, reptiles and amphibians, but are lower for fish (0.14), insects (0.10) and are lowest in birds (0.05). Values in plants range from 0.5 in selfing species to 0.1 in wind-pollinated outbreeding species. Typically F_{ST} values above 0.15 are considered to be an indication of significant differentiation among fragments.

Gene flow among population fragments

Gene flow reduces the genetic effects of population fragmentation

In a fragmented population with fixed and equal fragment sizes (N_e) and equal rates of gene flow (m) among all populations, an equilibrium between drift and migration is reached where F_{ST} is predicted to be:

$$F_{ST} = \frac{1}{(4N_e \, m + 1)} \tag{4.13}$$

A single migrant per generation is considered sufficient to prevent the complete differentiation of idealized populations, irrespective of their size

Sewall Wright obtained the surprising result that a single migrant per generation among idealized populations is sufficient to prevent complete differentiation (and fixation). Populations with migration rates of more than one migrant per generation exhibit no fixations, while those with less than one migrant per generation differentiate to the extent that some populations are fixed for alternative alleles.

These results are independent of population size. One migrant has as much impact on the equilibrium in a large population as in a small population. This appears paradoxical until it is recognized that one migrant represents proportionally a much higher migration rate in smaller than in larger populations. The higher migration rates in smaller populations counteract their greater loss of variation due to drift.

The conclusions above assume that migrants and residents are equally likely to produce offspring successfully, and that all population fragments have idealized structures, apart from the occurrence of gene flow. These assumptions are unlikely to be realistic. In real populations up to 10 migrants per generation may be required.

Population structure

The genetic impacts of population fragmentation may range from insignificant to severe, depending upon the details of the resulting population structures and migration patterns among fragments. In general, the overall long-term genetic consequences of different population structures are worst for **metapopulations** than for other structures, as they experience frequent bottlenecks associated with regular extinction and recolonization events. For example, there are about 1600 suitable meadows for the Glanville fritillary butterfly population in Finland, 320–524 being occupied in 1993–1996. In this metapopulation the population turnover rate is high, with an average of 200 extinctions and 114 colonizations per year, and deleterious genetic effects have been documented.

> The impact of population fragmentation depends on the details of the resulting population structure

Selection in small populations

In closed populations allele frequencies change predominantly through drift and selection. In large populations selection typically dominates allele frequency changes for alleles subject to natural selection. Conversely, in small populations genetic drift is usually the dominant force causing allele frequency change, even for alleles subject to natural selection. This is evident in the red flour beetle example in Fig. 4.2.

> Selection is less effective in small than large populations

The ability of drift to overpower selection depends on the selective advantage or disadvantage of the allele (i.e. the selection coefficient, s). When $s < 1/2N_e$, drift overpowers selection. For example, in populations with effective sizes below 50 (N below about 500), alleles with $s \leq 0.01$ are effectively neutral and subject to the effects of drift. This provides a critical insight for conservation genetics; selection is less effective in small than in large populations.

Glanville fritillary butterflies

SUGGESTED FURTHER READING

Frankham, R., J. D. Ballou & D. A. Briscoe. 2002. *Introduction to Conservation Genetics.* Cambridge University Press, Cambridge, UK. Chapters 8, 10, 11 and 13 have extended treatments of these topics, along with references.

Falconer, D. S. & T. F. C. Mackay. 1996. *Introduction to Quantitative Genetics*, 4th edn. Longman, Harlow, UK Chapter 3 provides a very clear introduction to the topics in this chapter.

Hanski, I. & M. Gilpin. (eds.) 1997. *Metapopulation Biology: Ecology, Genetics and Evolution.* Academic Press, San Diego, CA. A fine collection of relevant papers on fragmented populations. See especially the chapters by Hedrick & Gilpin, Barton & Whitlock and Giles & Goudet.

Hedrick, P. W. 2000. *Genetics of Populations*, 2nd edn. Jones & Bartlett, Boston, MA. Chapters 5–7 provide clear treatments of many of the genetic issues relating to small populations and population fragmentation.

Chapter 5

Genetics and extinction

Terms

Balancing selection, catastrophes, demographic stochasticity, environmental stochasticity, extinction vortex, genetic stochasticity, heterozygote advantage, inbreeding depression, lethal equivalents, major histocompatibility complex (MHC), minimum viable population size (MVP), 'mutational meltdown', polyploidy, population viability analysis (PVA), purging, self-incompatibility

Inbreeding and loss of genetic diversity are unavoidable in small populations of threatened species. They reduce reproduction and survival in the short term, diminish the capacity of populations to evolve in response to environmental change in the long term, and increase extinction risk

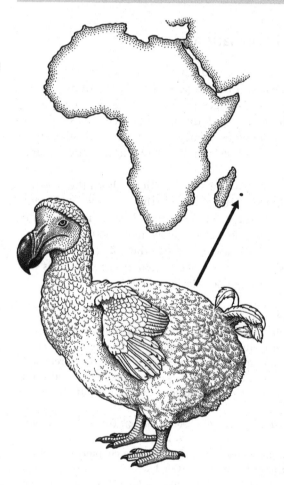

Extinct dodo and its previous distribution on the island of Mauritius

Genetics and the fate of endangered species

Little more than a decade ago, the contribution of genetic factors to the fate of endangered species was considered to be minor. Demographic and environmental fluctuations (stochasticity) and catastrophes were considered likely to cause extinction before genetic deterioration became a serious threat to wild populations. A healthy controversy has persisted. However, there is now a compelling body of both theoretical and experimental evidence indicating that genetic changes in small populations are intimately involved with their fate. Specifically, it has been shown that:

- many surviving populations are genetically compromised (reduced genetic diversity and inbred)
- loss of genetic diversity increases the susceptibility of populations to extinction, in at least some cases
- loss of genetic diversity is related to reduced fitness
- inbreeding causes extinctions in deliberately inbred experimental populations
- inbreeding contributes to extinctions in some natural populations and there is circumstantial evidence to implicate it in many other cases
- computer projections indicate that inbreeding will cause elevated extinction risks in threatened natural populations.

Inbreeding depression

Despite earlier scepticism, there is now clear and irrefutable evidence for inbreeding depression in both captive and wild populations. For example, inbred progeny had higher mortality than outbred progeny in 41 of 44 mammal captive populations (Fig. 5.1). In the case of the

The most immediate genetic threat to endangered species is inbreeding depression

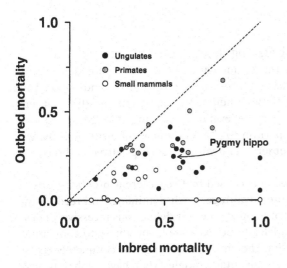

Fig. 5.1 Inbreeding depression for juvenile survival in 44 captive mammal populations (Ralls & Ballou 1983). Juvenile mortality in outbred individuals is plotted against that in inbred individuals from the same populations. The line represents equal survival of inbred and outbred individuals. *Most populations fall below the line, indicating that inbreeding is deleterious.*

Pygmy hippopotamus

pygmy hippopotamus, inbred offspring had 55% juvenile mortality, while outbred offspring had 25% mortality.

There is now also substantial evidence that inbreeding has deleterious consequences in the wild. For example, 90% of 157 data sets for 34 taxa showed inbred progeny with poorer survival, reproduction, or size, etc. than outbred progeny (Chapter 4).

Characteristics of inbreeding depression

To understand the probable impacts of inbreeding depression on wild populations, it is important to consider its characteristics and causation. These are:

- inbreeding depression has been observed in virtually every naturally outbreeding species that has been adequately investigated, and is to be expected in unstudied species
- natural outbreeding species in all major taxa show relatively similar average levels of inbreeding depression
- all components of reproductive fitness are subject to inbreeding depression
- characters most closely related to reproductive fitness show greater inbreeding depression than those that are peripherally related to fitness
- inbreeding depression is greater for total fitness than for its components
- inbreeding depression is greater in more stressful conditions
- families, populations and species differ in the extent of inbreeding depression
- inbreeding depression is proportional to the amount of inbreeding
- slower inbreeding generally causes less inbreeding depression than an equivalent amount of rapid inbreeding, but the difference is often small
- species that naturally inbreed show inbreeding depression, but the magnitude is generally less than that found in naturally outbreeding species.

Genetic basis of inbreeding depression

The magnitude of inbreeding depression depends upon the frequencies of deleterious alleles, the dominance of these alleles, the number of loci with deleterious alleles, and the amount of inbreeding

Inbreeding depression only occurs when there is dominance or heterozygote advantage and accumulates over all polymorphic loci affecting a trait

Inbreeding increases the frequency of homozygotes for deleterious alleles and the resulting inbreeding depression depends upon the frequencies of these alleles, their levels of heterozygosity, on their being recessive, and upon the level of inbreeding. Inbreeding reduces heterozygosity in proportion to the amount of inbreeding. Thus, we expect and generally observe a linear relationship between inbreeding depression and F.

Inbreeding depression is caused by two different processes, dominance and **heterozygote advantage**. With dominance, inbreeding increases the frequency of homozygotes for deleterious recessive alleles, thus allowing these alleles to be expressed. For alleles to contribute to inbreeding depression they must be partially or completely recessive. With heterozygote advantage, alleles that have higher fitness as heterozygotes cause inbreeding depression when inbreeding decreases the frequency of these fit heterozygotes relative to homozygotes. Which mechanism is operating is an important issue. While both processes cause inbreeding depression, they respond differently to natural selection. With dominance of advantageous alleles, selection can reduce the frequency of deleterious alleles (termed **purging**), but it cannot do this with heterozygote advantage. Dominance of favourable alleles is considered to make the major contribution to inbreeding depression.

Numerous loci (perhaps thousands) will generally be involved in causing inbreeding depression for fitness and its components.

Variability in inbreeding depression

Since inbreeding depression depends on the frequency of deleterious alleles, it is expected to have a large stochastic element. Small inbreeding populations are subject to genetic drift, and the same deleterious allele may be absent in one population and present at a relatively high frequency in another. Furthermore, individuals with the same expected F will have a range of actual levels of heterozygosity due to the sampling involved in Mendelian inheritance. Different loci will be homozygous in different individuals, just by chance. Consequently, different species, populations within a species, families within populations, and individuals within families will vary in their complement of deleterious alleles, and differ in their manifestation of inbreeding depression (Fig. 5.1). Since many loci affect reproductive fitness, it is highly improbable that fixation of deleterious alleles will be avoided at all loci. This accounts for the ubiquitous, but highly variable, nature of inbreeding depression.

> Inbreeding depression has a large stochastic element

However, most studies find little evidence of difference among major diploid taxa in the average extent of inbreeding depression for naturally outbreeding species. For example, there are no significant differences in inbreeding depression under captive conditions among mammalian orders.

Inbreeding depression is higher for gymnosperms than angiosperms. This may be related to a higher level of **polyploidy** (more than two doses of each chromosome) in the latter than the former. Since the rate of increase in homozygosity is slower in polyploids than in diploids, polyploids are expected to suffer less inbreeding depression.

Purging

Inbreeding increases homozygosity, exposing deleterious recessive alleles to selection. Consequently, deleterious alleles are removed, or **purged**, more rapidly than would occur under random mating.

> Inbreeding depression may be reduced, or purged, by selection against deleterious alleles, but this is unlikely to eliminate it

In populations with long histories of inbreeding, such as selfing plants, or animal populations that have remained small for a long time, the frequency of deleterious alleles may have been reduced by purging. This raises the question: do we expect to see less inbreeding depression in these?

Naturally inbreeding species and populations that have historically been reduced in size still show inbreeding depression, but the magnitude is generally less than in other populations

While plant species that self-fertilize or naturally inbreed still show levels of inbreeding depression, the magnitude is usually less than in related outbreeding populations. This implies that some purging has occurred. Studies also suggest that a history of small population size in animals reduces but does not eliminate inbreeding depression.

Although there is considerable variation in the extent of inbreeding depression, its almost universal occurrence indicates that wildlife managers must adopt strategies to minimize inbreeding in threatened populations.

Measuring inbreeding depression

Inbreeding depression is usually measured as the proportionate decline in the mean of a quantitative character per unit increase in inbreeding coefficient

Inbreeding depression is often measured by comparing the fitnesses of inbred organisms with non-inbred control cohorts. Control cohorts should be contemporary with the inbreds and have experienced the same environmental conditions.

There are two basic approaches to measuring inbreeding depression:

- taking the ratio of inbred mean to non-inbred mean, or
- regressing the measure of a character (e.g. survival, fecundity, size) against inbreeding coefficients obtained from pedigrees.

A general measure of inbreeding depression (δ) is the proportionate decline in mean due to a given amount of inbreeding:

$$\delta = 1 - \frac{\text{fitness of inbred offspring}}{\text{fitness of outbred offspring}} \qquad (5.1)$$

This formula does not specify the level of inbreeding and this must be defined for each case. Example 5.1 illustrates the use of equation 5.1 to estimate inbreeding depression in Dorcas gazelle.

δ is most often used in plants. Since many plants can be selfed, the usual estimate of inbreeding depression is obtained by comparing selfed and outcrossed progeny; the impact of inbreeding due to an inbreeding coefficient of 50%.

Example 5.1 | Inbreeding depression in Dorcas gazelle

Juvenile survival of 50 outbred and 42 inbred Dorcas gazelle were 72.0% and 40.5%. Any individual with $F > 0$ was defined as inbred, but the average level of inbreeding in them was approximately 0.25 (full-sibs). The inbreeding depression (δ) for juvenile survival in this species is:

$$\delta = 1 - \frac{\text{fitness of inbred offspring}}{\text{fitness of outbred offspring}} = 1 - \frac{0.405}{0.720} = 0.44$$

Dorcas gazelle

Lethal equivalents

The usual means for expressing and comparing the extent of inbreeding depression for survival in animals is **lethal equivalents**. This is the slope of the regression of survival on level of inbreeding. The probability of surviving to a particular age, S, can be expressed as:

$$S = e^{-(A+BF)} \tag{5.2}$$

where e^{-A} is fitness in the outbred population, F is the inbreeding coefficient, and B is the rate at which fitness declines with increasing inbreeding.

If we take natural logarithms (ln) this becomes a linear relationship:

$$\ln S = -A - BF \tag{5.3}$$

B measures the additional genetic damage that would be expressed in totally homozygous individuals ($F = 1$). Thus, B is the number of lethal equivalents per gamete, and $2B$ the number per individual. One lethal equivalent is defined as a group of detrimental alleles that would cause on average one death if homozygous, e.g. one lethal allele, or two alleles each with 50% probability of causing death, etc.

To estimate lethal equivalents, data are collected on survival rates of individuals with different inbreeding coefficients and weighted

F	Lived	Died
0	86	55
0.125	5	2
0.25	12	18
0.375	1	5

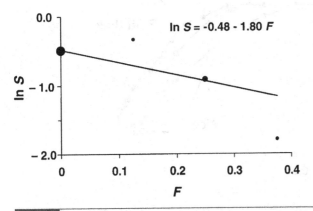

Okapi

Fig. 5.2 Relationships between survival and inbreeding coefficient in okapi (data from de Bois *et al.* 1990). The natural logarithm of survival is plotted against the inbreeding coefficient and the regression line is inserted. The sizes of the circles indicate sample sizes.

linear regression is used to estimate A and B. Figure 5.2 illustrates the relationship between ln S and inbreeding coefficient for the okapi, with the regression line inserted. The slope of the line (B) is -1.80, indicating that the population contains 1.8 haploid and 3.6 diploid lethal equivalents. Ralls and co-workers found that the median number of lethal equivalents for 40 captive mammal populations was 1.57 (B) per haploid and 3.14 ($2B$) per diploid, although species varied widely. These values indicate that each gamete contains deleterious mutations equivalent to between one and two lethals when homozygous. Similar values have been calculated for other animal populations, including humans.

Relationship between inbreeding and extinction

Deliberately inbred populations of laboratory and domestic animals and plants have greatly elevated extinction rates

The first evidence on the relationship between inbreeding and extinction came from deliberately inbred populations of laboratory and domestic animals and plants. Between 80% and 100% of deliberately inbred populations die out after eight generations of brother–sister mating or three generations of self-fertilisation. For example, all 338 populations of Japanese quail, inbred by continued brother–sister mating, were extinct after four generations. Examples from mice and fruit flies are shown in Fig. 5.3.

Rate of inbreeding and extinction risk

Even slow inbreeding increases the risk of extinction

Natural populations of outbreeding animals and plants are usually subject to slow rates of inbreeding, dependent on their population sizes. Slower inbreeding allows natural selection more generations to purge deleterious alleles and generally leads to less inbreeding depression than rapid inbreeding for the same total amount of inbreeding. However, differences are typically small.

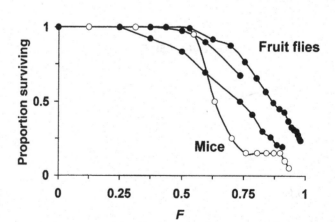

Fig. 5.3 Relationship between inbreeding and extinction (Frankham 1995). Populations of mice and two species of fruit flies (one with two populations) were inbred using brother–sister matings. Demographic stochasticity made very little or no contribution to these extinctions. *The proportion of populations going extinct rises with inbreeding, but extinctions do not begin until intermediate levels of inbreeding have been reached.*

Even slow rates of inbreeding increase the risk of extinction; it just takes longer for inbreeding to accumulate and extinction to occur. For example, 15 of 60 fruit fly populations, inbred due to sizes of 67 individuals per generation, went extinct within 210 generations, while five of six replicate housefly populations of size 50 went extinct over 64 generations.

Inbreeding and extinction in wild populations

Since inbreeding leads to elevated extinction risks in captive populations, it is logical to extrapolate this to wild populations. However, linking extinction in wild populations to genetic effects is difficult, as determining exact causes of extinction in the wild is never easy and necessary genetic data are often lacking. Nevertheless, three lines of evidence support this contention:

> There is growing evidence that inbreeding elevates extinction risks in wild populations

- many surviving wild populations suffer loss of genetic diversity and inbreeding depression
- inbreeding and loss of genetic variation have been shown to contribute to extinction of populations in nature
- computer projections predict that inbreeding will increase extinction risks for wild populations.

Many wild populations are genetically compromised

Some conservation biologists have predicted that demographic and environmental stochasticity and catastrophes would usually drive populations to extinction before genetic factors become a problem. This prediction is not generally true, as the majority of threatened species have lower genetic diversity than related non-threatened species and thus are likely to be suffering inbreeding depression and loss of adaptability. For example, black-footed rock wallabies on Barrow Island and other islands off the west coast of Australia have very low levels of genetic diversity and the only island population investigated (Barrow Island) exhibits inbreeding depression (Box 5.1). In addition, euros (a larger kangaroo species) on Barrow Island also have reduced genetic diversity and suffer from chronic anaemia and higher parasite loads than their mainland counterparts.

> Most threatened species have reduced genetic diversity

Box 5.1	Island populations of black-footed rock wallabies have persisted for more than 1600 generations at small sizes, are highly inbred, have low levels of genetic diversity and exhibit inbreeding depression (Eldridge *et al.* 1999)

Rock wallabies are 1m tall kangaroos inhabiting rocky outcrops on the Australian mainland and offshore islands (see map below). The Barrow Island population of black-footed rock wallabies (location 1) has been isolated from the mainland for 8000 years (about 1600 generations) and has a relatively small population size.

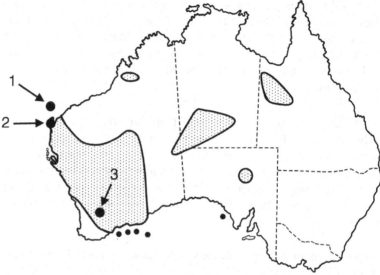

Black-footed rock wallaby

Microsatellite diversity in the Barrow Island population, and in other island populations (dots without numbers) is markedly lower than in mainland sites at Exmouth (2) and Wheatbelt (3).

Population (location)	Proportion of loci polymorphic	Mean number of alleles per locus	Average heterozygosity
Barrow Island (1)	0.1	1.2	0.05
Mainland			
Exmouth (2)	1.0	3.4	0.62
Wheatbelt (3)	1.0	4.4	0.56

Since its isolation, the Barrow Island population has obviously survived stochastic fluctuations and catastrophes, but suffers genetic problems that increase its risk of extinction. It has an inbreeding coefficient of 0.91 and displays inbreeding depression compared to mainland populations. The frequency of lactating females is 92% in mainland rock wallabies, but only 52% on Barrow Island.

Island populations have been viewed as ideal sources for restocking depleted or extinct mainland populations, especially in Australasia. However, they are poor candidates for translocations if alternative mainland populations still exist, as they typically have low genetic diversity and are inbred.

Direct and circumstantial evidence of extinctions due to inbreeding and loss of genetic diversity

There is direct evidence that inbreeding and loss of genetic diversity increase the risk of extinction for populations in nature

Direct evidence for inbreeding being involved in extinction of natural populations has been presented for Finnish butterfly populations (Box 5.2). Inbreeding was a significant predictor of extinction risk after the effects of all other ecological and demographic variables had been accounted for. Further, experimental populations of the evening primrose plant founded with a low level of genetic diversity (and high

Box 5.2 Inbreeding and extinction risk in butterfly populations in Finland (Saccheri et al. 1998)

Using molecular genetic analyses, levels of heterozygosity were determined in 42 butterfly populations in Finland in 1995, and their extinction or survival recorded in the following year. Of these populations, 35 survived to autumn 1996 and seven went extinct. Extinction rates were higher for populations with lower heterozygosity, an indicator of inbreeding, even after accounting for the effects of demographic and environmental variables (population size, time trend in population size and area) known to affect extinction risk, as shown in the figure below. The different curves represent the relationships between extinction probability and proportion of loci heterozygous for populations with different numbers (1–5) of larval groups.

Glanville fritillary butterflies

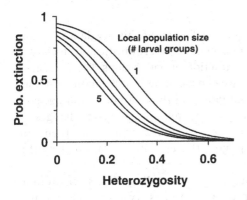

Bighorn sheep

inbreeding) exhibited 75% extinction rates over three generations in the wild, while populations with lower inbreeding showed only a 21% extinction rate.

Smaller populations are expected to be more prone to extinction than larger ones for demographic, ecological and genetic reasons. A strong relationship was observed between population size and persistence in North American bighorn sheep (Fig. 5.4). All populations

Small populations are more likely to suffer from extinctions than large populations for both genetic and ecological reasons

Fig. 5.4 Extinction rates are higher in smaller than in larger populations. Relationship between persistence and population size in North American bighorn sheep (after Berger 1990).

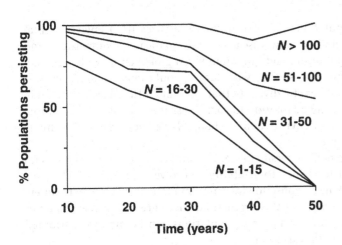

<50 became extinct within 50 years. A wide variety of demographic, ecological and genetic (including inbreeding and loss of genetic diversity) factors may have contributed to these extinctions. Similarly, mammalian extinctions in national parks in western North America were related to park area, and presumably population sizes. Extinctions were more frequent for populations with smaller initial population sizes, larger fluctuations in population size and shorter generation times. While these effects are expected from demographic and environmental considerations, they are also predicted from genetic theory.

Declines in population size or extinction in the wild have been attributed, at least in part, to inbreeding in many other populations including Florida panthers, Isle Royale gray wolves, greater prairie chickens, heath hens, middle spotted woodpeckers, adders, and many island species.

Computer projections

Computer projections predict that inbreeding elevates extinction risk for most outbreeding wild populations

Computer projections incorporating factual life history data are often used to assess the combined impact of all deterministic and stochastic factors on the probability of extinction of populations. This is termed **population viability analysis** (see later). In brief, information on population size, birth and survival rates and their variation over age and years, together with measures of inbreeding depression, changes in habitat quality, etc. form the input. Stochastic models are then run through repeated cycles to project the fate of populations into the future.

Brook and colleagues found that inbreeding effects on juvenile survival at levels found in captive mammals reduced median times to extinction by an average of 25–31% across a range of 20 animal species at population sizes typical for threatened species. A related computer projection for a rare European gentian plant yielded similar predictions.

Relationship between loss of genetic diversity and extinction

Populations with lower genetic diversity are poorer at coping with environmental extremes and diseases than populations with higher genetic diversity

Natural populations face continuous pressures from environmental changes including new diseases, pests, parasites, competitors and predators, pollution, and human-induced global climate change. Naturally outbreeding species with large populations normally possess large stores of genetic diversity that generate differences among individuals, allowing adaptation to such pressures (Chapter 2). Evolutionary responses to environmental change have been observed in many species (Chapter 3).

Loss of genetic diversity in small populations of threatened species is predicted to reduce their ability to evolve, and increase their extinction risk in response to environmental change. While experimental evidence validates this prediction, there are only a few examples where extinctions of natural populations can be directly attributed to lack of genetic variation.

Relationship between loss of genetic diversity at self-incompatibility loci and extinction in plants

The most direct evidence of a relationship between loss of genetic diversity and increased risk of extinction comes from studies of **self-incompatibility** loci in plants. About half of all flowering plant species have genetic systems to prevent self-fertilization. Self-incompatibility is regulated by one or more loci that may carry 50 or more alleles in large populations. If the same allele is present in a pollen grain and the stigma, fertilization by that pollen grain will not be successful. Self-incompatability is presumed to have evolved to avoid the deleterious effects of inbreeding.

Self-incompatibility alleles may be lost by genetic drift in small populations, leading to a reduced proportion of pollen that can fertilize the eggs of any individual, and eventually to reduced seed set and extinction. For example, the Illinois lakeside daisy population declined to three plants and did not reproduce for 15 years despite bee pollination. It had so few self-incompatibility alleles that it was functionally extinct. Plants did however produce viable seed when fertilized with pollen from large populations in Ohio and Canada. The endangered grassland daisy in eastern Australia also exhibits reduced fitness in smaller populations with reduced numbers of self-incompatibility alleles. While reduced fitness due to loss of self-incompatibility alleles has been documented in only a few species of plants, it is likely to become a problem in most threatened, self-incompatible plants.

> Loss of self-incompatibility alleles in small populations of many plant species lead to reduced reproductive fitness

Endangered grassland daisy

Relationship between loss of genetic diversity and susceptibility to diseases, pests and parasites

Novel pathogens constitute one of the most dire threats to all species. Loss of genetic diversity severely diminishes the capacity of populations to respond to this pressure. For example, the American chestnut was driven near to extinction in the 1950s by the introduced chestnut blight disease, as it had no genetic variation for resistance. Previously, the chestnut had dominated the northeastern forests of the USA and this event constitutes one of the largest ecological disasters to strike that country.

The **major histocompatibility complex** (MHC) is a large cluster of loci in vertebrates involved in recognizing foreign antigen molecules from pathogens and regulating immune responses. Genetic variation at these loci is amongst the highest known for any loci. The more heterozygous an individual is, the more pathogens the individual can respond against. Similarly, the more variable a population is, the more likely it is that at least some individuals can resist a pathogen. This genetic diversity is maintained by selection favouring heterozygotes and by selection favouring rarer alleles (**balancing selection**). It is therefore predicted, and supported by some evidence that loss of diversity at these loci reduces the ability of populations to respond against novel forms of pathogens. Despite selection maintaining diversity,

> Populations with low genetic diversity are expected to suffer more seriously from diseases, pests and parasites than those with high genetic diversity

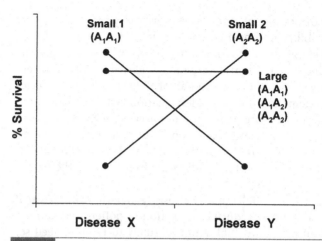

Fig. 5.5 Hypothetical example of the relationship between genetic diversity and disease resistance. Two small inbred populations, each homozygous for different alleles, are resistant to one disease but not to the other. Conversely, a larger population containing both alleles has resistance to both diseases.

Desert topminnow fish with trematode parasites (black spots)

alleles will be lost by genetic drift in small populations, increasing the chance that a pathogen that can kill one individual can kill all. Following sequential assaults by different pathogens, populations with high genetic diversity are more likely to persist than populations with low genetic diversity (Fig. 5.5). For example, topminnow fish populations with low genetic diversity had higher parasite loads than populations with greater diversity. Mortality due to gastrointestinal nematode parasites, following periods of high population density, is elevated in Soay sheep with low genetic diversity and high inbreeding.

Genetically viable populations

Shortage of space for threatened species

As resources for maintaining threatened species are limited, it is critical to identify the minimum population size needed to retain genetic 'health', by avoiding inbreeding depression and retaining evolutionary potential. Current population sizes of threatened species are typically too small to avoid genetic deterioration

There is a severe shortage of habitat in the wild for many threatened species, and a critical lack of facilities for captive breeding. It is therefore essential to establish how these resources are best used. This section addresses the question: 'How large must populations be, to be genetically viable in the long term?' This issue has been discussed under the title of **minimum viable population size** (MVP), the minimum size required to retain reproductive fitness and evolutionary potential over thousands of years. This does not signify that populations of lesser size have no future, only that their reproductive fitness and evolutionary potential are likely to be compromised, and that they will have an increased risk of extinction. As Soulé noted: 'There are no hopeless cases, only people without hope and expensive cases.'

How large?

Three genetic components must be considered in answering this question:

- Is the population size large enough to avoid inbreeding depression?
- Is there sufficient genetic diversity for evolution to occur in response to environmental change?
- Is the population large enough to avoid accumulating new deleterious mutations?

Different authors have derived various estimates (Table 5.1). We consider each of these issues below.

Table 5.1 How large must populations be to retain genetic 'health'? Various estimates of the required effective population size (N_e) are given. The times to recover normal levels of genetic diversity following complete loss of diversity are also given

Goal	N_e	Recovery time (generations)
Avoid inbreeding depression	50	
Retain evolutionary potential	500	$10^2–10^3$
	5000	
	570–1250	
Retain single-locus genetic diversity	$10^5–10^6$	$10^5–10^7$
Avoid accumulating deleterious mutations	12–1000	

Sources: see Frankham *et al.* (2002).

Retaining reproductive fitness

Franklin and Soulé both suggested that an effective population size of 50 was sufficient to avoid inbreeding depression in the short term, based on the experience of animal breeders. However, this is likely to be a severe underestimate for the longer term.

No finite population is immune from eventual inbreeding depression

Since inbreeding increases at a rate of $1/(2N_e)$ per generation, all finite closed populations eventually become inbred. Further, as inbreeding depression is proportional to the inbreeding coefficient, there is no known threshold below which inbreeding is not deleterious. Even low levels of inbreeding are expected to result in some low level of inbreeding depression. Based upon the impacts of inbreeding found in captive mammals, we would expect about 2% inbreeding depression when $F = 0.01$, 4% when $F = 0.02$, and 15% when $F = 0.10$ for juvenile survival alone.

Inbreeding depression has been described in fruit fly populations maintained for many generations at effective sizes of about 50 and in housefly populations of $N_e \sim 90$ within five generations.

Populations with effective sizes of 50 in fruit flies and 90 in houseflies show inbreeding depression

We do not know precisely how large populations must be to avoid meaningful inbreeding depression in the long term, but the required size is clearly much greater than an effective size of 50. Disturbingly, about one half of all captive populations of threatened mammals have N of less than 50, and are likely to be suffering inbreeding depression, or are likely to do so relatively soon.

At what point will inbreeding become sufficient to cause extinctions? Estimated times to extinction for different sized housefly populations approximated the effective size in generations, i.e. 54 generations for $N_e = 50$. Extinction risks in rapidly inbred populations of mice and fruit flies increase markedly at $F = 0.5$ and beyond (Fig. 5.3 above).

In practice, wild populations that were listed as endangered in 1985–1991 numbered 100–1000 individuals. Similarly, the IUCN scheme for categorization of extinction risk lists 50, 250 and 1000 adults in populations of stable sizes as cut-offs for the critically endangered, endangered and vulnerable categories. Since N_e/N ratios are about 0.1, many of these populations will have effective sizes of 50 or less and are at risk of extinction from inbreeding depression unless their sizes are substantially increased.

Retaining evolutionary potential

Effective population sizes of 500–5000 have been suggested as necessary to maintain evolutionary potential

Since our goal is to conserve species as dynamic entities capable of responding to environmental change, evolutionary potential must be retained. There is a range of estimates of the population size required to maintain long-term genetic diversity and evolutionary potential, but general agreement that it is an N_e of at least 500 (Table 5.1). Since this debate has major implications for the practical genetic management of wild and captive populations, we explore the basis of the estimates.

Franklin predicted an effective size of 500, arguing that additive genetic variation determined evolutionary potential, and this is directly related to heterozygosity. He assumed that the level of additive genetic variation at equilibrium was dependent upon a balance between loss due to drift and replenishment by mutation. Using information on the mutation rate for a quantitative character, he concluded that an effective size of 500 was required to maintain the original level of additive genetic variation.

Lande suggested that a value of 5000 was required, arguing that only about 10% of newly generated mutations are useful for future genetic change, as most are deleterious. While reservations have been expressed about this estimate, it is generally accepted that an effective size between 500 and 5000 is required to retain short-term evolutionary potential.

Wild populations in nature require adult census sizes about 10 times larger than the N_e values estimated above, i.e. several thousand to tens of thousands

Census sizes in wild populations must be about one order of magnitude higher than the N_e values we have calculated, i.e. 5000–50 000, since comprehensive estimates of N_e/N are about 0.1 (Chapter 4). This sets a lower limit for the minimum size to maintain long-term viability, and falls within the range of values reached from consideration of other non-genetic threats.

The fate of species with $N_e < 500$

Species with effective sizes insufficient for long-term maintenance of genetic diversity are not doomed to immediate extinction. On average they will suffer depletion of genetic diversity and reduced ability to evolve. They will become inbred at a rate dependent on their size, with consequent reduction in reproduction and survival rates, and require increasing human intervention to ensure their survival. This may take the form of environmental enhancement (isolating them from competitors, avoiding introduction of diseases, habitat restoration, etc.), or managing them to increase reproduction and survival.

Species with effective sizes of less than 500 are not doomed to immediate extinction, but will become increasingly vulnerable with time

Retaining single locus genetic diversity in the long term

Some loci, such as self-incompatibility loci in plants and MHC loci in vertebrates, are so important to survival that it is essential to retain their genetic diversity. The population sizes required to retain this diversity are much larger than those for quantitative characters, as mutation rates are low for individual loci. Based on mutation–drift equilibrium, Lande & Barrowclough suggested effective population sizes of 10^5–10^6 were required. These sizes are unattainable for most species of conservation concern (especially vertebrates), given current habitat availability and conservation resources. Indeed stable populations of this size are not even considered as threatened. Population sizes required to maintain diversity at loci subject to balancing selection (e.g. SI and MHC loci) will be less than this estimate, but may also be unattainable goals.

Effective population sizes of 10^5–10^6 are required to retain single-locus diversity due to the balance between mutation and drift

Time to regenerate genetic diversity

Loss of genetic diversity would be of little concern if it were regenerated rapidly by mutation. However, mutation rates are very low, so regeneration times are very long (Table 5.1). Single-locus diversity with a mutation rate of 10^{-5}–10^{-7} per generation takes 100 000 to 10 million generations to regenerate. Quantitative genetic variation requires only 100–1000 generations to regenerate – still about 2600–26 000 years for elephants! Since we cannot rely on mutation for regeneration in time spans of conservation concern, every effort must be made to preserve extant genetic diversity.

If genetic diversity is lost, it can only regenerate very slowly by mutation, with recovery to original levels taking hundreds or thousands of generations

Avoiding accumulation of new deleterious mutations

Alleles that only have small beneficial or detrimental effects on a trait become effectively neutral in small populations. That is, their fate is determined by chance rather than their selection coefficients. Thus, a proportion of new mildly deleterious mutations will become homozygous in small populations, resulting in reduced reproductive fitness that may eventually lead to extinction ('**mutational meltdown**').

Some mildly deleterious mutations are fixed by chance in small populations and result in reduced reproductive fitness. The size of populations required to avoid such effects is unclear

The significance of mutational accumulation in sexually reproducing populations is controversial. Estimates of effective population sizes required to prevent declines due to mutational accumulation have ranged from 12 to 1000.

Genetic goals in the management of wild populations

Few management programs for endangered species in the wild include genetic objectives

We are aware of only a few management plans for endangered species in the wild where genetic objectives are defined. In the golden lion tamarin, the objective is to retain 98% of genetic diversity for 100 years, corresponding to $N_e \sim 400$. Currently the census size is about 600 wild individuals, plus over 400 animals in the forests resulting from reintroductions and about 500 individuals in captivity. The N_e/N ratio must exceed 0.27 to attain this goal based on all animals, or 0.4 for wild animals. Since this is unlikely, the genetic goal is not being achieved, primarily as a consequence of lack of available habitat for population expansion.

Genetic goals in the management of captive populations: a compromise

Captive populations of endangered species are usually managed to retain 90% of their genetic diversity for 100 years

As discussed previously, there are many fewer resources available than would be required to maintain all the species deserving captive breeding, especially if the numbers recommended above are used (e.g. $N_e = 500$ per species). Zoos house about 540 000 vertebrates and at best, only half of the spaces are suitable for propagating endangered animals. It is estimated that about 2000 vertebrate species require captive breeding to prevent extinction. To maintain each of these species at an effective size of 500 (assuming $N_e/N = 0.3$ in captivity) requires 3.3 million animal spaces, about 12 times that available.

The current compromise is to manage captive endangered species to conserve 90% of the extant genetic diversity for 100 years. The 100 years time frame derives from the hope that wild habitat may become available following the predicted human population decline in 100–200 years. Based upon equation 4.2, this requires an N_e of:

$$N_e = \frac{475}{L} \tag{5.4}$$

where L is the generation length in years. Consequently, the required size is inversely proportional to generation length, one of the few circumstances where long-lived species are at an advantage. For example, the effective size required to maintain 90% of the original heterozygosity is 1769 for the white-footed mouse with a generation length of 14 weeks, 475 for a species with one generation per year, and 18 for Caribbean flamingos with a generation time of 26 years.

While maintaining 90% of genetic diversity for 100 years may be a reasonable practical compromise, it is unlikely that all the species requiring captive breeding can be housed. Further, the cost of this compromise is increased inbreeding and reduced reproductive fitness. An accepted 10% loss of heterozygosity corresponds to an increased inbreeding coefficient of 10%, with consequent inbreeding depression. After 100 years, captive-bred individuals will be related to each other somewhere between that of first cousins ($F = 0.0625$) and half-siblings ($F = 0.125$), reducing juvenile survival by about 15% and total fitness by about 25%. The fitness costs are likely to be much greater if species are subsequently reintroduced into harsher wild environments. Thus, captive breeding programs are balancing the risks of moderate inbreeding over 100 years against the benefits of maintaining additional endangered species within the available resources. This is a necessary, but not attractive choice.

Population viability analysis (PVA)

Genetic impairment is only a part of the threatening processes faced by species. Wild populations face a range of threats from both deterministic and stochastic factors that may act, and interact, to drive populations to extinction. Consequently, genetic concerns must be considered in the broader conservation context. This section provides such a connection and is concerned with assessing extinction risk from all factors. We then consider means for evaluating management options for restoring threatened and endangered species.

> Population viability analysis assesses the combined impacts of deterministic and stochastic factors on extinction risk, allowing evaluation of alternative management options in species recovery programs

Assessments of extinction risk are required so that populations can be ranked according to relative risk, and conservation priorities set. Numerical evaluation of risks can be provided through population viability analysis (PVA). Here, we present a brief introduction to the field.

Deterministic factors

Deterministic factors are those processes that have a consistent direction and a relatively consistent magnitude. Most of the deterministic factors that cause decline and extinction are directly, or indirectly associated with human actions, namely:

> A large and increasing number of species have been reduced by deterministic factors associated directly, or indirectly, with human actions

- destruction of habitat for urban and agricultural development, etc.
- over-exploitation for commercial or recreational use
- inadvertent pollution and deliberate application of pesticides, herbicides, etc.
- exotic species introduced intentionally, or by accident (e.g. ballast water, international trade)
- combinations of the above factors.

While habitat loss is the primary documented factor, several of the factors combine to drive species towards extinction in most cases. Further, these deterministic factors reduce population sizes to the point where additional stochastic processes may become significant, and deliver the final blow.

Stochastic factors

Unlike deterministic factors, stochastic processes in small populations have large random components with effects varying in direction and

> Small populations face additional threats: demographic, environmental and genetic stochasticity, and catastrophes

magnitude. As briefly outlined in Chapter 1, there are four forms of stochasticity relevant to extinction risk in small populations:

- **demographic stochasticity** This describes the natural fluctuations in birth and death rates and sex-ratios. Extinction can result if, by chance, all individuals in a small population are sterile, or all of one sex. For example, the last six dusky seaside sparrows were all males, an event with a probability of $(\frac{1}{2})^6 = 1.6\%$.
- **environmental stochasticity** Birth and death rates may vary due to variation in the environment, such as fluctuations in rainfall, temperature, density of competitors, predators, food sources, etc. For example, birth and death rates are strongly affected by rainfall in red kangaroos.
- **genetic stochasticity** This encompasses inbreeding depression, loss of genetic variation and accumulation of new deleterious mutations.
- **catastrophes** Extreme environmental events such as cyclones, severe winters, fires, floods, volcanic eruptions and disease epidemics may be the final cause of extinctions. For example, a hurricane caused a significant decline in population numbers of the endangered Puerto Rican parrot, while African lions in the Serengeti recently suffered high mortality due to canine distemper, and many frog species throughout the world are being driven to extinction by a fungal disease.

Puerto Rican parrot

> Stochastic factors operate in a feedback cycle termed the 'extinction vortex'

Interactions of stochastic factors

The combined impacts of stochastic factors are more damaging than the sum of their individual effects. Human pressures typically lead to small population sizes. This promotes inbreeding and consequent reductions in birth and survival rates. In turn, this causes further reductions in population size, increased demographic instability and a downward cycle to extinction, termed the 'extinction vortex' (Fig. 5.6).

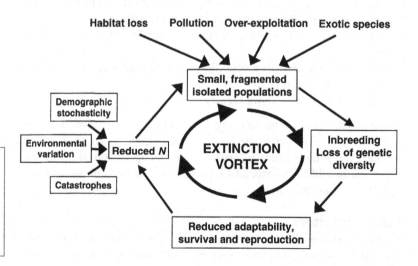

Fig. 5.6 The extinction vortex. This describes the possible interactions between human impacts, inbreeding, loss of genetic diversity and demographic instability in a downward spiral towards extinction.

Variation in population size due to demographic and environmental stochasticity and catastrophes reduces the effective population size and increases the rate of inbreeding, and thus increases extinction risk.

Combined impacts

The total threat experienced by a population is the combined effect of deterministic factors, and demographic, environmental and genetic stochasticity, plus the occasional catastrophe. Consequently, actions to recover threatened species must not only address the original causes of decline (usually deterministic factors), but also cope with the additional stochastic threats. Identifying the most important factors determining extinction risk can help identify possible remedial action for threatened populations. Population viability analysis (Fig. 5.7) is used to predict extinction risk from the combined impacts of all deterministic and stochastic threats.

> Extinction risk reflects the combined impacts, and interactions of all deterministic and stochastic factors

Predicting extinction probabilities: population viability analysis (PVA)

PVAs are usually carried out by inputting, to a computer program, information on

- birth and survival rates and their variances
- number of populations
- population sizes
- habitat capacities
- frequencies and effects of threats (e.g. catastrophes, hunting, etc.)
- other details about species life history (e.g. susceptibility to inbreeding depression, rates of gene flow between populations, etc.).

The populations are then projected forward in time (Fig. 5.8). The concepts used in the computer simulation programs are based upon the accumulated knowledge of more than 100 years of research into

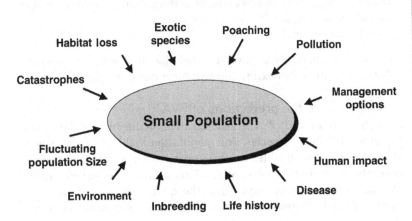

Population viability analysis (PVA)

Habitat loss · Exotic species · Poaching · Pollution

Catastrophes

Small Population

Management options

Fluctuating population Size

Environment · Inbreeding · Life history · Disease · Human impact

Fig. 5.7 Population viability analysis (PVA) models the effects of different life history, environmental, and threat factors on the population size and extinction risk of populations or species.

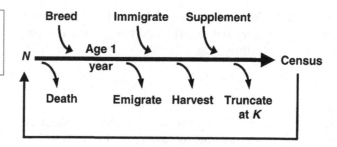

Fig. 5.8 Cycle of events in a typical population viability analysis (PVA) run as it progresses through generations.

population demography, ecology and genetics. An example of the kind of input information required to run a PVA for the software package VORTEX is given for the golden lion tamarin in Fig. 5.9.

PVAs are used to predict extinction risk in a particular species, to compare alternative management options for recovering a threatened species, and as a research tool. When used to compare recovery options, they may assess the impacts of control on poaching, or the impacts of removing a predator, etc.

Many replicate runs (typically 500–1000) are conducted for a given set of input data, as individual population projections will vary amongst these stochastic simulations. For example, the variability among replicate runs for the Capricorn silvereye bird (all using identical input data) is shown in Fig. 5.10. Results are usually summarized over all simulations. PVAs usually report population sizes, population growth rate, and proportion of simulations extinct, and some report proportion of heterozygosity retained.

Replicate runs of PVA software using the same input parameters give widely varying population trajectories as a consequence of stochastic variation

Genetics and PVA

PVAs show that inbreeding depression often increases the risk of extinction in threatened species

Inbreeding depression is the only genetic factor that has been incorporated into PVAs. For this incorporation, we need to know:

- susceptibility of the species to inbreeding depression
- what fitness components are affected by inbreeding (survival, fecundity)
- the extent of isolation among fragments, i.e. the migration and gene flow rates, as these affect the inbreeding coefficient
- the breeding system (outbreeding vs. selfing, monogamous vs. polygamous vs. hermaphrodite, etc.); this affects N_e
- population size and sex-ratio.

As discussed above, PVAs have revealed that inbreeding depression substantially increases extinction risk for a range of species.

How useful are the predictions of PVA?

PVA often has its greatest value as a tool to assist planning for the recovery of threatened species, to allow adaptive management, to determine sensitivities and to compare recovery options, rather than in providing accurate predictions of extinction risk

A major limitation of PVA is that insufficient life-history data exist for most threatened species. Full population viability analyses may therefore not be possible, or they may have low reliability. However, the most important contributions of risk assessments using PVA do not necessarily come from the quantitative assessments of extinction risk themselves. Rather, the process of conducting a PVA involves:

TAMARIN.OUT	***Output Filename***		
N	***Graphing Files?***	12.5	***Adult Fmort***
1000	***Simulations***	6.7	***EV–AdultFemaleMortality***
100	***Years***	29.8	***MMort age 0***
I	***Reporting Interval***	7.0	***EV–MaleMortality***
0	***Definition of Extinction***	19.7	***MMort age I***
I	***Populations***	9.9	***EV–MaleMortality***
Y	***Inbreeding Depression?***	23.9	***MMort age 2***
3.14	***Lethal equivalents***	5.0	***EV–MaleMortality***
50.0	***Prop. genetic load lethals***	17.2	***MMort age 3***
N	***EV concordance repro – surv?***	7.5	***EV–MaleMortality***
4	***Types Of Catastrophes***	16.0	***Adult Mmort***
L	***Monogamous, Polygynous, Hermaphroditic, or Long-term Monogamous***	8.2	***EV–AdultMaleMortality***
		7.0	***Prob. of Catastrophe I***
		1.0	***Severity–Reproduction***
4	***Female Breeding Age***	0.9	***Severity–Survival***
4	***Male Breeding Age***	1.0	***Prob. of Catastrophe 2***
16	***Maximum Age***	1.0	***Severity–Reproduction***
50	***Sex Ratio***	0.5	***Severity–Survival***
5	***Max Litter Size (0 = N distribn)*****	33.33	***Probability of Catastrophe 3***
N	***Density Dependent Breeding?***	1.0	***Severity–Reproduction***
75.7	***breeding***	0.99	***Severity–Survival***
3.82	***EV–Reproduction***	5.0	***Probability of Catastrophe 4***
20.77	***Population I:% Litter Size I***	1.0	***Severity–Reproduction***
57.94	***Population I:% Litter Size 2***	0.95	***Severity–Survival***
5.42	***Population I:% Litter Size 3***	Y	***All Males Breeders?***
15.21	***Population I:% Litter Size 4***	Y	***Start At Stable Age Distribution?***
32.8	***FMort age 0***	350	***Initial Population Size***
8.7	***EV–FemaleMortality***	350	***K***
19.6	***FMort age I***	0.0	***EV–K***
13.3	***EV–FemaleMortality***	N	***Trend In K?***
24.6	***FMort age 2***	N	***Harvest?***
7.6	***EV–FemaleMortality***	N	***Supplement?***
21.0	***FMort age 3***	N	***AnotherSimulation?***
5.0	***EV–FemaleMortality***		

Fig. 5.9 Example of input information for a population viability analysis (PVA) run using the VORTEX software package for the golden lion tamarin (after Ballou et al. 1998). This is only for one of several populations of this species. EV is the environmental variation for the parameter. Input variables are explained in the VORTEX software manual.

- summarizing information about the life history of the species
- identifying all the threatening processes impacting upon it
- assessing their likely importance (sensitivity analysis)
- identifying potential recovery strategies and evaluating their relative impacts
- identifying deficiencies in knowledge about the species, and formulating research proposals to remedy them.

Thus, considerable benefits may be gained through the PVA process, even if the quantitative predictions are not particularly accurate. PVA

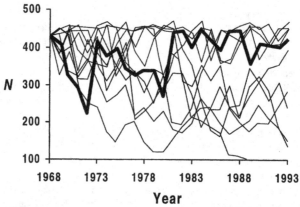

Capricorn silvereye

Fig. 5.10 Stochastic variability among replicate PVA runs in projections for the Capricorn silvereye population on Heron Island, Great Barrier Reef, Australia (from Brook & Kikkawa 1998). Input data are identical for all runs; the variability observed results from the computer program using random number generators to mimic the demographic, environmental and catastrophic stochasticity natural to populations. How much variability is added by the random number generator depends on the population size, sampling theory and the variance of rates provided by the input data. The bold line is the observed size of the silvereye population.

provides a transparent planning process that should have internal consistency. Further, the recovery process can be operated in an adaptive manner. PVA projections can be updated as more information on the species is gathered and management practices modulated in light of PVA predictions.

Minimum viable population sizes (MVP)

As habitat and financial resources are limited, it is essential that we determine the minimum sizes and habitat areas required to maintain viable populations in the long term. PVA was originally devised to determine minimum viable population sizes (MVP) and habitat areas for grizzly bears.

There is a consensus that the size required for a population to be viable in the long term is at least thousands to tens of thousands

Different estimates of the size required, based on a variety of theoretical arguments, and on empirical data are given in Table 5.2. Based on empirical evidence, Thomas suggested that 10 is far too small, 100 is usually inadequate, 1000 is adequate for species with typical variability in population sizes, while 10 000 should permit medium to long-term persistence of birds and mammals that show major fluctuations in population size. The required size is not universal and is widely assumed to be strongly dependent on details of the biology and environment of the species and the types of threats faced. However, this assumption may be an artefact of measuring it for a fixed number of years, as the MVPs predicted for 99% persistence for 40 generations were relatively consistent across major taxa, and environments in a study of 100 species.

Table 5.2 | Sizes required for long-term viability of populations to overcome different threats. Variation refers to the propensity for population sizes to fluctuate

	N_e	N
Threat		
1. Loss of genetic diversity	500–5000	5000–50 000
2. Mutational accumulation	1000	10 000
3. Demographic stochasticity		10s–100
4. Environmental stochasticity		1000+
5. Catastrophes		1000+
Empirical data (Thomas 1990)		
Birds and mammals:		
average variation		1000
high variation		10 000
Insects		
average variation		10 000
high variation		100 000
PVAs for 100 vertebrate species		>6 000

Sources: Thomas (1990); Nunney & Campbell (1993); Reed et al. (2003).

From Table 5.2 it is clear that populations must number at least a few thousand to be viable in the long term. Both the population size at the time species are listed as threatened and the recovery targets under the USA Endangered Species Act are typically too small. The median size at listing is about 1000 individuals for animals and 100 for plants. Further, the median population size for a taxon to be considered recovered was about 1550.

Population sizes used to list and delist threatened species are usually less than those recommended above

A worrying implication of these numbers is that even the largest reserves (apart from the Antarctic) are too small to maintain adequate population sizes for long-term survival of large herbivores and especially large carnivores.

SUGGESTED FURTHER READING

Frankham, R., J. D. Ballou & D. A. Briscoe. 2002. *Introduction to Conservation Genetics.* Cambridge University Press, Cambridge, UK. Chapters 2, 12, 14 and 20 have extended treatments of these topics, along with references.

Beissinger, S. R. & D. R. McCullough. 2002. *Population Viability Analysis.* University of Chicago Press, Chicago, IL. Proceedings of a conference on PVA. See especially contributions by Mills & Lindberg, Ralls et al. and Shaffer et al.

Ecological Bulletin. 2000. Volume 48: *Population Viability Analysis.* Special issue on PVA. See especially papers by Akçakaya & Sjögren-Gulve, Akçakaya, and Lacy.

Ralls, K., J. D. Ballou & A. Templeton. 1988. Estimates of lethal equivalents and the cost of inbreeding in mammals. *Conservation Biology* 2, 185–193. Classic paper on the impacts of inbreeding on captive mammals.

Saccheri, I., M. Kuussaari, M. Kankare, P. Vikman, W. Fortelius & I. Hanski. 1998. Inbreeding and extinction in a butterfly metapopulation. *Nature 392*, 491–494. Describes the first direct evidence that inbreeding contributes to the extinction of wild populations in nature.

Westemeier, R. L., J. D. Brawn, S. A. Simpson, T. L. Esker, R. W. Jansen, J. W. Walk, E. L. Kershner, J. L. Bouzart & K. N. Paige. 1998. Tracking the long-term decline and recovery of an isolated population. *Science 282*, 1695–1698. Describes the decline of a small, isolated greater prairie chicken population in Illinois due to loss of genetic diversity and inbreeding, and its recovery following introduction of unrelated birds from other states.

Young, A. G., A. H. D. Brown, B. G. Murray, P. H. Thrall & C. H. Miller. 2000. Genetic erosion, restricted mating and reduced viability in fragmented populations of the endangered grassland herb *Rutidosis leptorrhynchoides*. Pp. 335–359 in A. G. Young & G. M. Clarke, eds. *Genetics, Demography and Viability of Fragmented Populations*. Cambridge University Press, Cambridge, UK. Describes fitness consequences of loss of genetic diversity for self-incompatibility alleles in an endangered plant.

Chapter 6

Resolving taxonomic uncertainties and defining management units

Taxonomic status must be accurately established so that endangered species are not denied protection, nor effort wasted on abundant species. Genetic information assists in resolving taxonomic uncertainties and defining management units within species

Terms

Allopatric, allopolyploid, autopolyploid, biological species concept, cryptic species, ecological exchangeability, evolutionarily significant units (ESU), exchangeability, genetic distance, lineage sorting, management unit, outbreeding depression, phylogenetic trees, polyploidy, sibling species, speciation, sub-species, sympatric, taxa

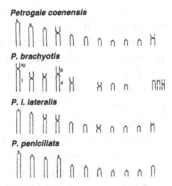

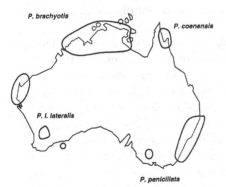

Rock wallabies in Australia, along with the chromosomes from a sample of species. Many are endangered. Genetic methods have helped resolve taxonomic uncertainties in this group.

Importance of accurate taxonomy in conservation biology

The taxonomic status of many taxa is unresolved. This is particularly true for lower plants and invertebrate animals, but also applies to large and obvious animals including deer, wallabies and wolves.

In conservation, many erroneous decisions may result if the taxonomic status of populations is not correctly assigned. These include:

- unrecognized endangered species may be allowed to become extinct
- endangered species may be denied legal protection while populations of common species, or hybrids between species, may be granted protection
- incorrectly diagnosed species may be hybridized with other species, resulting in reduced reproductive fitness
- resources may be wasted on abundant species, or hybrid populations
- populations that could be used to improve the fitness of inbred populations may be overlooked.

Velvet worm

This chapter explores the rationale and methodologies for defining taxonomic status. A similar approach is used to distinguish management units within species, as the crossing of genetically differentiated populations can result in reduced population fitness (**outbreeding depression**) and disruption of unique evolutionary groups.

Taxonomic uncertainties result predominantly from inadequate data. Many species' descriptions trace to limited information on the geographic distribution of a small number of (usually morphological) traits of unknown genetic basis. Velvet worms (Phylum Onychophora) in Australia provide an extreme example. Only seven named species were recognized in a 1985 review, based on morphology. However, over 100 clearly diverged species have now been identified using allozymes and microsatellites.

Incorrect 'lumping' of several distinct species into one recognized species has denied protection to endangered species. The threatened tuataras in New Zealand, the only surviving members of an ancient reptilian order, are now known to include two species, one of which was at serious risk of extinction (Box 6.1). Likewise, endangered Kemp's Ridley sea turtle is a genetically distinct species, rather than a form of a related non-endangered species, as was previously thought. It is now being afforded protection. *Helianthus exilis*, a sunflower from California, was denied protection as it was closely related to another sunflower with which it hybridized. However, molecular genetic analyses identified *H. exilis* as a distinct endangered species deserving conservation.

Box 6.1 | **Taxonomic uncertainty in tuataras and North American pumas and their conservation implications** (Daugherty *et al.* 1990; Culver *et al.* 2000)

Tuatara

TUATARA

The threatened tuataras in New Zealand are the only survivors of an ancient reptilian order, and were thought to be a single species. Studies of different island populations, using both 25 allozyme loci and morphology, revealed that it consists of three distinct groups, *Sphenodon punctatus punctatus* in the north, a western Cook Straits sub-species (*S. p. western*), and *S. guntheri*. The latter was being neglected and at serious risk of extinction without active conservation management.

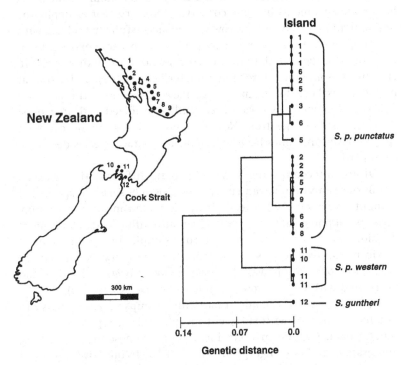

NORTH AMERICAN PUMAS

Mammalogists have recognized approximately eight morphological sub-species of North American pumas (also referred to as cougars, panthers and mountain lions), including the critically endangered Florida panther. However, microsatellite and mtDNA analyses found no significant differentiation among the populations, but did separate these from South American sub-species.

An important conservation implication is that the number of separate sub-species requiring conservation is reduced. Further, the recent controversial decision to augment Florida panthers with individuals from the Texas sub-species, to remove

Florida panther

inbreeding depression, can now be recognized as a logical management action involving only simple translocation of animals, rather than a controversial decision to hybridize distinct sub-species.

Conversely, 'splitting' of one species into two or more recognized taxa may lead to erroneous conservation decisions. Two rare flightless chafer beetles in New Zealand, previously considered to be different species, have been shown to belong to a single species, based on allozyme data. Similarly, there is no genetic differentiation among the eight recognized morphological sub-species of North American pumas, including the critically endangered Florida panther (Box 6.1). The distinctiveness of the northern and California spotted owls in western North America is highly controversial.

Hybridization between populations whose taxonomy is incorrect has created problems in some conservation efforts. For example, the sole remaining dusky sea sparrow was unsuccessfully hybridized with an inappropriate sea sparrow sub-species and became extinct. Infertility in a captive population of dik-dik was due to mixing of different chromosome races (probably undescribed species), and a similar problem has been found in owl monkeys. Conversely, the last Norfolk Island boobook owl was successfully hybridized with its closest relative, a sub-species from New Zealand. Hybrids among common species have sometimes been mis-identified as rare species deserving conservation.

Failure to recognize the degree of genetic differentiation between the Bornean and Sumatran sub-species (possibly separate species) of orangutans previously led to their hybridization in many zoos. Concerns about potential hybridization have also influenced management decisions in the opposite direction. For example, two populations of Mexican wolf were kept separate from the one known small and inbred 'pure' population, as it was incorrectly suspected that they had hybridized with dogs, coyotes or gray wolves (see Chapter 9).

In practice, the taxonomy of particular groups of populations can usually be resolved with sufficient morphological, reproductive and genetic data. Genetic markers including chromosomes, allozymes, microsatellites, DNA fingerprints and mtDNA frequently help. However, to appreciate this use of genetic markers, we must first review what is meant by a biological species, what we seek to conserve and how populations differentiate and speciate.

What is a species?

There is no universally accepted definition of species

Most named species have been delineated, on the basis of morphological characteristics, as groups of individuals that are distinct from all other groups. However, morphological definitions of species may have limited connection to genetics or evolution. Some groups of individuals initially appear morphologically indistinguishable, but are composed of two or more distinct species (**cryptic** or **sibling species**). Chinese and Indian muntjac deer are morphologically similar, yet the former has 46 chromosomes, while the latter has 6 in males and 7 in females, and they are clearly distinct species.

Confusingly, there is no universally accepted definition to answer the question 'What is a species?' At least 22 definitions exist. These include definitions based on morphology, ecology and genetics to definitions based on biological characteristics, evolutionary histories and phylogeny. Some definitions can, and have, classified the sexes of one species as separate species. Other more useful definitions are based on evolutionary units and gene flow.

The **biological species concept** has been the most influential definition of species in population and evolutionary genetics and in conservation biology. This defines a species as a group of actually, or potentially, interbreeding individuals and natural populations that cannot interbreed with individuals from all other such groups. This definition recognizes that individuals within a species can exchange genetic material, while those from different species normally do not.

> Definitions of species generally recognize that individuals within an outbreeding species can exchange genetic material with each other, but not with individuals from other species

This definition provides practical means for delineating species genetically. Groups of individuals sharing the same area will be exchanging genetic material if they belong to the same species, but not if they belong to separate species. Geographically separated populations of the same species will be capable of crossing and producing fertile offspring in the first and subsequent generations. Conversely, populations of different species will either fail to mate, or produce offspring with reduced survival or fertility. For example, lions and tigers can be hybridized, but their progeny are sterile.

The biological species concept does not deal adequately with asexual and habitually inbreeding forms, becomes blurred for species that hybridize and has little relevance to classification of fossil specimens. Given these limitations, it is not surprising that the biological species concept is controversial. The US Endangered Species Act is based on the biological species concept, but it has encountered difficulties by excluding hybrids from conservation and by not dealing adequately with asexual forms.

Lack of a universally recognized definition of species creates enormous difficulties in conservation biology.

Sub-species

Threatened **sub-species** are frequently accorded legislative protection, and are the focus of substantial conservation effort. Sub-species are groupings of populations, within a species, that share a unique geographic range or habitat and are distinguishable from other subdivisions of the species in several genetically based traits. Members of different sub-species do not normally exhibit marked reproductive isolation. They can usually produce fertile offspring, although there may be some reduction in fertility or survival of these offspring. Crosses

> Sub-species are partially differentiated populations within a species

between the Bornean and Sumatran sub-species of orangutans produce fertile offspring with no apparent reduction in survival rates.

The concept of sub-species is more subjective than that of species. They may best be considered as populations partway through the evolutionary process of divergence towards full speciation.

Distinct populations within species and sub-species may also be accorded legislative protection, as described later in the chapter.

How do species arise?

Speciation involves the genetic divergence of populations until they are reproductively isolated

Species arise in two ways. The first is diversification, when a prior species gives rise to two or more descendant species. This occurs when populations genetically differentiate, become reproductively isolated and are said to have speciated. Speciation frequently involves at least partial physical isolation. The second is gradual change within a species over time so that it is considered to be a different species at a later time. In this chapter, we are concerned with the diversification of species.

Isolating factors

Populations may become isolated by geographic features (allopatry), or a change (e.g. host shifts) within the same environment (sympatry)

Physical isolating factors often result from geographic changes (mountain building, desertification, river diversion, sea-level changes and continental drift) or from spread of organisms to novel territories. If the isolated populations become so different that they do not interbreed upon secondary contact then speciation is called **allopatric**. This is a generally considered to be the most common form of speciation in animals.

Speciation may also occur within the range of the ancestral species, such as when a species spreads from one host to another. This is termed **sympatric** speciation. For example, hawthorn flies that court and mate on the developing fruits of their host plants began to utilize apples in 1864 and cherries in 1960. As these trees have somewhat different fruiting times, this provided an isolating mechanism and selective forces drive differentiation among populations on the different hosts. This may be a common form of speciation in parasites. Evidence for the importance of sympatric speciation is accumulating.

Reproductive isolation arises from adaptation to different environments

Recent evidence indicates that adaptation to distinct environments has an important, or even predominant, role in leading to reproductive isolation. For example, three-spine stickleback fish in three isolated lake populations in western Canada independently evolved bottom-dwelling and open-water forms with different sizes and diets, following glacial retreat 10 000 years ago. Bottom-dwelling forms from different lakes are not reproductively isolated, nor are open-water forms from different lakes. However, bottom-dwelling and open-water forms from the same or different lakes show reproductive isolation.

Many plant species have been formed 'instantly' due to polyploidy

'Instant' speciation

Many plant species have arisen 'instantly' via **polyploidy** when the chromosome number increases, e.g. $4n$ (tetraploid). These forms are, to a large degree, reproductively isolated from their progenitors. For example, the giant California redwood tree is a hexaploid with 66 chromosomes while its closest related living relative has $2n = 22$.

Two forms of polyploidy occur, **autopolyploidy** and **allopolyploidy**. Autopolyploids form by increasing the number of sets of chromosomes from within a species, presumably by production of diploid gametes. For example, there are both diploid ($2n = 22$) and autotetraploid ($4n = 44$) forms of the endangered grassland daisy in eastern Australia.

Allopolyploid species form by combining the complete chromosomal constitutions from two pre-existing species, as described for the rare allotetraploid Hong Kong lady's tresses orchid in Fig. 6.1. This form of polyploidy is much commoner in nature than autopolyploidy.

The frequency of polyploidy varies among plant taxa, being 47–70% in angiosperms, 95% in ferns, but only 5% in gymnosperms. Polyploidy is much rarer in animals (especially mammals and birds) than in plants. Sexually reproducing polyploids are known in a few groups of fish and frogs, but polyploid animals are more frequently parthenogenetic (reproduction from an unfertilized egg), for example

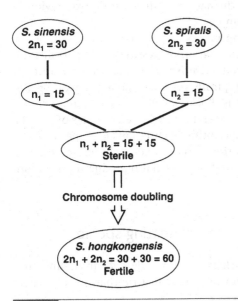

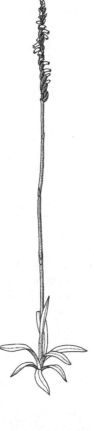

Fig. 6.1 'Instant' speciation in the rare Hong Kong lady's tresses orchid through allopolyploidy (after Sun 1996). The rare orchid *Spiranthes hongkongensis* arose by combining the full chromosome complements of two distinct species, each with 30 chromosomes (*S. sinensis* and *S. spiralis*), yielding a species with 60 chromosomes.

The probable mode of formation for *S. hongkongensis* is indicated above. The initial cross between the diploid progenitor species yielded a sterile hybrid. A spontaneous doubling of chromosome numbers, presumably in a flower bud (or production of rare diploid gametes) generated gametes that yielded a fertile allopolyploid. Evidence from allozymes indicates that allotetraploid *S. hongkongensis* formed only once as almost all individuals have the same multilocus genotype. This orchid can cross with its diploid progenitor *S. sinensis*, but the progeny are infertile triploids.

some amphibians, reptiles and insects. Polyploid animals of conservation concern include many species of tetraploid salmonid fish.

Speciation is generally slow

Speciation usually takes thousands to millions of years, apart from that due to polyploidy

Most speciation occurs gradually over long periods of time, presumed to be thousands to millions of years. For example, some plant populations that have been geographically isolated for at least 20 million years, such as sycamores and plantains in America and Asia, still form fertile hybrids. Similarly, some birds distributed in both Europe and America, such as tits, creepers and ravens, are so similar that they are classified as the same species. Some cases of speciation have been more rapid, but still involve many thousands of years. For example, some populations of polar bears and voles separated by Pleistocene glaciation 1.8 million years ago have developed at least partial reproductive isolation. On average, fruit flies (the best-studied group) take from 1.5–3.5 million years of separation to speciate. However, Hawaiian fruit flies have speciated in as little as 500 000 years. Cichlid fish in Lake Nabugabo, Africa have speciated within 4000 years. About 170 species of cichlids evolved in Lake Victoria within 500 000–750 000 years. They are now being exterminated by an introduced carnivorous fish. Sympatric speciation in the hawthorn flies mentioned previously is one of the most rapid examples.

As evolution is an ongoing process, some populations will be observed partway through the speciation process. They are partially differentiated, show some reproductive isolation and are very difficult to classify. Sub-species may typically represent populations progressing towards full species status. Rock wallabies in Australia (chapter frontispiece), many of which are endangered, are examples of populations 'caught in the act' of speciating. Populations and species show varying degrees of differentiation in morphology, chromosomes, allozymes and mtDNA. Many exhibit only partial reproductive isolation, and there are several hybrid zones.

Use of genetic analyses in delineating species

Analyses using genetic markers can usually aid in delineating species, but this is more definitive for sympatric than for allopatric populations.

Use of genetic analyses in delineation of sympatric species

Genetic analyses can be used to provide a definitive diagnosis of the taxonomic status of sympatric populations

Sympatric populations share the same, or overlapping, geographic distributions. According to the biological species concept, sympatric populations of the same species should exchange alleles, while different species sharing the same geographic region should not. Consequently, if any genetic marker shows lack of gene exchange, two sympatric populations belonging to different species have been identified. In practice, several loci are required for such diagnoses. For example, two sympatric forms of potoroos (small marsupials) in southeastern

Long-footed potoroo

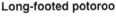

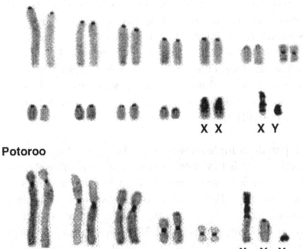

X X   X Y

Potoroo

X Y₂ Y₁

Long-footed potoroo

Fig. 6.2 Two sympatric potoroos in southeastern Australia belong to separate species, as indicated by lack of gene exchange (Seebeck & Johnson 1980). Potoroos are small nocturnal marsupials, akin to pint-sized kangaroos. Between 1967 and 1978, four potoroos were collected in southeastern Australia that appeared to differ from the potoroo species known to occur in that area. The long-footed potoroo has longer hind feet, is about twice the size of the other sympatric potoroos, and the two forms have different chromosome numbers and do not share alleles at five allozyme loci.

Australia were shown to be different species based on their different chromosome numbers and the lack of shared alleles at five allozyme loci (Fig. 6.2). Clearly, the two forms are reproductively isolated. The long-footed potoroo exists in very low numbers and is an endangered species. Morphologically similar velvet worms from the same log in the Blue Mountains, Sydney, Australia had fixed differences at 86% of 21 allozyme loci. Consequently, they are clearly different species that diverged millions of years ago.

If two sympatric populations share alleles at all loci, then the hypothesis that they are the same species cannot be rejected. For example, two rare, different-coloured sympatric chafer beetles from New Zealand were found to belong to the same species based on allozymes and morphology.

Use of genetic analyses in delineation of allopatric species

As allopatric populations are physically isolated, they typically have no opportunity for exchange of genetic material. The biological species concept would ideally require evidence from crosses between the populations to distinguish species. Hybrid sterility, or markedly reduced survival, would indicate separate species. Conversely, if the hybrids were fully viable and fertile through several generations, then the populations would belong to the same species. Such crosses are usually impractical, especially in threatened species.

Allopatric populations that differ chromosomally are normally distinctive species. However, the use of other genetic markers is less definitive, and requires calibration against genetic differentiation of other recognized species

Consequently, genetic markers are often used to delineate allopatric species. Fixed chromosomal differences normally provide definitive evidence for distinct species status, as many chromosomal differences result in partial sterility in heterozygous individuals. For example, Chinese and Indian muntjac deer are morphologically similar, but are clearly distinct species as they have different chromosome numbers, as described above. Similarity of chromosomes in allopatric populations is not definitive evidence that they belong to the same species, but differences in chromosomes indicate distinct species.

Classification based on molecular markers is more arbitrary than for sympatric species, as it is based on inferred reproductive isolation. In practice, two populations are considered to be different species if they are as genetically differentiated as are two well-recognized species in a related group. For example, Bornean and Sumatran orangutans differ in mtDNA, proteins and DNA fingerprints, and by a chromosomal inversion, and differ as much as do other well-known distinct primate species. Consequently, it has been suggested that they be classified as different species, as opposed to two sub-species (Box 6.2). However, crosses between the two forms are viable and fertile in the F_1 and F_2 generations.

Genetic markers have been used to establish that newly discovered forms of mammals in Vietnam and Laos are distinct from known species. For example, the saola or Vu Quang bovid has been identified as a distinct species based on morphology and DNA analyses. Only a few hundred individuals of this species are presumed to exist, so it must be considered endangered.

| **Box 6.2** | Genetic differentiation between Bornean and Sumatran orangutans: are they separate species? (Xu & Arnason 1996; Zhi et al. 1996) |

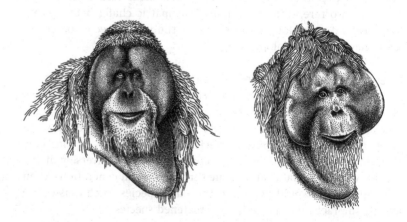

Bornean and Sumatran orangutans are restricted to their respective islands in Southeast Asia. They differ in morphology and behaviour, and have been designated as sub-species. Bornean and Sumatran orangutans differ by a pericentric chromosomal inversion (a reversed chromosomal segment), in mtDNA sequences (below) in protein coding nuclear loci (below) and, DNA fingerprints. Estimates of time since divergence from a common ancestor average 1.7 million years, far more ancient than the separation of Borneo and Sumatra by rising sea levels 10–20 000 years ago. As they differ genetically by at least as much as do the clearly recognized chimpanzees and bonobos, full species status for the two forms has been suggested.

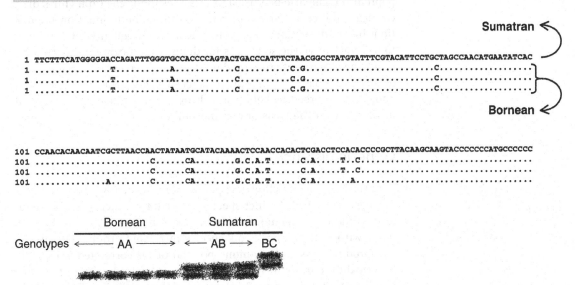

Hybrids are found in many zoos and are viable and fertile in the F_1 and F_2 generations. Consequently, they cannot be recognized as distinct species according to the biological species concept, but they can be so recognized under other definitions of species. This illustrates the confusion created by different species definitions. The proposal to classify them as separate species has been challenged. Regardless of their formal taxonomic status, they represent distinct evolutionary units and should be managed separately.

If differentiation between allopatric populations is much less than that between two well-recognized species in the same or related genera, then the populations are considered to belong to the same species. For example, the colonial pocket gopher population from Georgia, USA consisted of fewer than 100 individuals in the 1960s and was listed as an endangered species. Subsequent analyses based on morphology, allozymes, chromosomes and mtDNA revealed no consistent differences between this population and nearby populations of the common southeastern pocket gopher. On this basis, the colonial pocket gopher does not warrant recognition as a separate species.

There is only limited evidence on the ability of morphology, chromosomes, allozymes, nuclear DNA markers and mtDNA to correctly predict taxonomic status as determined by breeding experiments. Data on rock wallabies and native rodents in Australia suggest that chromosomes provide better predictions of reproductive isolation than molecular markers. Unfortunately, chromosomal analyses are currently out of fashion for delineating taxonomic status.

Mitochondrial DNA is one of the most frequently used genetic markers to delineate taxa. However, it has a number of limitations. First, mtDNA differentiation can be produced by lack of female dispersal, while male dispersal may be keeping populations genetically homogeneous at nuclear loci. Second, mtDNA patterns in different populations can also be misleading as a result of selection. Third, drift can lead to incorrect trees of relationship if the foundation population, before divergence, was polymorphic. Consequently, it is unwise to use mtDNA as the sole basis for delineating taxonomic status.

In the absence of crossing data, the most convincing delineations of species are based on the concordance of a wide array of information (morphology, breeding behaviour, chromosomes, nuclear markers and mtDNA), or on the basis of chromosomes.

Genetic distance

Nei's genetic distance is the most commonly used measure of genetic differentiation among populations and species

In order to delineate allopatric populations as merely populations or as sub-species or as distinct species, we require a measure of genetic differentiation, or **genetic distance**. We can then compare this distance with that among 'good' species in related groups. The extent of reproductive isolation among populations is correlated with their genetic differentiation (Table 6.1). The most commonly used measure is Nei's genetic distance, D_N. First we define Nei's index of genetic similarity, I_N

$$I_N = \frac{\sum\limits_{i=1,}^{m}(p_{ix}\,p_{iy})}{\sqrt{\left(\sum\limits_{i=1,}^{m}p_{ix}^2\right)\left(\sum\limits_{i=1,}^{m}p_{iy}^2\right)}} \qquad (6.1)$$

and then transform this to obtain Nei's genetic distance

$$D_N = -\ln(I_N) \qquad (6.2)$$

where p_{ix} is the frequency of allele i in population (or species) X, p_{iy} is the frequency of allele i in population (or species) Y, m is the number of alleles at the locus, and ln is logarithm to the base e.

When allele frequencies are similar in two populations ($p_{ix} = p_{iy}$), the genetic similarity approaches 1, and the genetic distance approaches zero. Conversely, when the two populations share no alleles, the index of genetic similarity is zero and the genetic distance is infinite. Example 6.1 illustrates calculation of genetic distance from an allozyme locus in endangered red-cockaded woodpeckers. Estimates should be based on numerous loci to provide reliable estimates of genetic distances.

Example 6.1 | Calculation of Nei's genetic distance

Two populations of endangered red-cockaded woodpeckers have the following frequencies at the lactate dehydrogenase (Ldh) locus (from Stangel *et al.* 1992).

Ldh allele	Frequencies	
	Vernon	Apalachicola
B	0.023	0.019
C	0.977	0.885
D	0.000	0.096

To calculate the genetic distance we need to compute the squares of the allele frequencies and the products of their frequencies between populations. For the Vernon population the sum of the squared frequencies is

$$\Sigma p_{ix}^2 = 0.023^2 + 0.977^2 + 0.000^2 = 0.955$$

and for Apalachicola

$$\Sigma p_{iy}^2 = 0.019^2 + 0.885^2 + 0.096^2 = 0.793$$

The numerator is the sum of the cross products, as follows:

$$\Sigma(p_{ix}\,p_{iy}) = 0.023 \times 0.019 + 0.977 \times 0.885 + 0 \times 0.096 = 0.865$$

Consequently, Nei's genetic similarity for the Vernon and Apalachicola comparison is

$$I_N = \frac{\Sigma(p_{ix}\,p_{iy})}{\sqrt{(\Sigma p_{ix}^2)\,(\Sigma p_{iy}^2)}} = \frac{0.865}{\sqrt{0.955 \times 0.793}} = \frac{0.865}{0.870} = 0.994$$

and the genetic distance is

$$D_N = -\ln(I_N) = -\ln(0.994) = 0.006$$

Consequently, the genetic distance between the Vernon and Apalachicola populations of the woodpeckers is only 0.006, i.e. the populations are genetically very similar.

How large are genetic distances for 'good' species?

Genetic distances generally increase as we progress up the taxonomic hierarchy from populations within species, to species, genera and families. For example, average allozyme genetic distances in fruit flies of the *Drosophila willistoni* species complex increases from geographically isolated populations (mean $D = 0.03$), to sub-species ($D = 0.23$), to distinct species ($D = 1.21$). However, the relationship is very approximate and 'noisy', as can be seen in Table 6.1. For example, sub-species of lizards show greater genetic distances than species of macaques, gophers and birds.

Genetic distances generally increase with level of reproductive isolation, but the relationship is very 'noisy'

Table 6.1 Genetic differences between sub-species and species, using Nei's genetic distance, D_N

Comparison	D_N
Sub-species	
Red deer	0.02
Mice	0.19
Pocket gophers	0.004–0.26
Ground squirrels	0.10
Lizards	0.34 –0.35
Drosophila willistoni group	0.23±0.03
Plants (peppers)	0.02–0.07
Species	
Macaques	0.02–0.10
Ground squirrels	0.56
Gophers	0.12
Birds (*Catharus*)	0.01–0.03
Galapagos finches	0.004–0.07
Lizards (*Anolis*)	1.32–1.75
Lizards (*Crotaphytus*)	0.12–0.27
Amphisbaenians	0.61–1.01
Salamanders	0.18–3.00
Teleosts (*Xiphophorus*)	0.36–0.52
Teleosts (*Hypentelium*)	0.09–0.33
Fruit flies (distinct species)	
obscura group	0.29–0.99
willistoni group	1.21±0.06
Hawaiian species	0.33–2.82
Plants (peppers)	0.05 –0.79

Source: Nei (1987).

Constructing phylogenetic trees

Information from genetic or morphological markers can be used to construct phylogenetic trees

We are often interested in evolutionary relationships among populations and species (see Chapter 9). For example, we may wish to identify the most closely related population, or sub-species to use in crossing programs to recover threatened species. For example, if the true relationships among the six sub-species of seaside sparrow had been known, the last of the dusky sub-species would have been crossed, probably successfully, to a related Atlantic coast sub-species rather than to a Gulf coast form. Phylogenetic trees reflecting evolutionary relationships among species (or populations) can be constructed using genetic data. Example 6.2 provides a simple illustration of tree building based on mtDNA sequence data for the Norfolk Island boobook owl and its nearest presumed relatives.

A large number of statistical methods are now available for deriving trees from molecular markers or from morphology, including distance matrix methods (UPGMA), maximum parsimony and maximum likelihood methods. Software packages are available for computation (e.g. PAUP, PHYLIP, HENNIG86). These methods generally yield reliable trees if there is sufficient information, i.e. numbers of loci, number of nucleotides or amino acids or number of morphological characters, or preferably a combination of these.

Example 6.2 | Building a phylogenetic tree from DNA sequence data (Norman et al. 1998)

mtDNA sequence differences in the Norfolk Island (NI) boobook owl and its nearest presumed relatives are given below. The NI boobook owl declined to a single individual, and the best recovery option was to cross it to its most closely related sub-species. Crosses of the remaining female to a male from a related New Zealand sub-species has yielded 16 F_1 offspring. Subsequent analyses of mtDNA sequences (298 bases) confirmed that the sub-species chosen for crossing was the most closely related and conformed with morphological data.

Norfolk Island boobook owl

Comparison	mtDNA base pair differences
Norfolk Island boobook – New Zealand boobook	2
Norfolk Island boobook – Tasmanian boobook	8
New Zealand boobook – Tasmanian boobook	8
Powerful owl – rufous owl	13
Norfolk Island boobook – powerful owl	21
Norfolk Island boobook – rufous owl	23

By placing the nearest relatives closest together, we obtain the following tree. The lengths of the segments are proportional to the number of base differences between taxa, and therefore should approximate the evolutionary time since divergence. Thus, the branch lengths to the node below NI boobook and NZ boobook are 1 each (total of 2 differences between them). These differ from Tas boobook by 8 bases, so we attribute 4 to the segment from Tas boobook to the node below it, and 4 from the node to NI boobook and NZ boobook. As one base difference has already been attributed to the distance to the node to NI boobook and NZ boobook, 3 are attributed to the section from their node to the joint node with Tas boobook. Half of the difference of 13 between powerful owl and rufous owl (6.5) is attributed to each path to their node. Finally the NI boobook – powerful owl and NI boobook – rufous owl average 22 bases different, so we attribute half of this (11) to each path to the lowest node. On the left-hand side we have already attributed 4, leaving 7 for the section to the lowest node. On the right-hand side, 6.5 has been allocated to the path to the first node, leaving 4.5 to allocated to the path from this node to the lowest node.

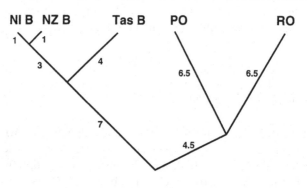

Trees will accurately reflect evolutionary relationships only if the markers are neutral, the rates of neutral mutation are similar in different lineages and the foundation population is monomorphic. For example, the mtDNA tree for primates disagrees with information from a wide array of morphological, chromosomal, behavioural and other genetic evidence, probably as a consequence of selection. When selection occurs or mutation rates are unequal in different lineages, rates of evolution differ, the lengths of branches are distorted, and the inferred tree may not agree with the real tree.

> Trees based on genetic markers will only accurately reflect evolutionary relationships if rates of evolution are constant in different lineages, markers are neutral and the foundation population is monomorphic (or there are many independently inherited markers)

When the starting population is polymorphic, as frequently is the case, fixation of different initial sequences in different lineages (termed **lineage sorting**) may lead to incorrect inferred phylogenies. This situation can be resolved if data are available from many independent loci. About 30 polymorphic allozyme loci or 20 microsatellite loci are recommended. The consensus phylogeny in primates, based on several nuclear loci, is concordant with the other evidence.

Phylogenetic trees, or gene trees, are used for many other purposes in conservation biology (Chapter 9).

Outbreeding depression

Crossing of genetically differentiated populations or higher taxa creates the risk that hybrid offspring in the first and subsequent generations will suffer reduced reproductive fitness.

> Outbreeding depression is the reduction in reproductive fitness resulting from crossing of populations

Many well-defined sub-species have diverged to the point where crosses among them result in reduced fitness. Crosses between Bengal and Siberian tigers would not be expected to produce offspring fit for either environment. A major reason for resolving taxonomic uncertainties is to avoid establishment of such crosses.

> Outbreeding depression may be expected in crosses between different sub-species or species

Populations that have differentiated as a result of adapting to different habitats may show reduced fitness when crossed because hybrids may be adapted to neither habitat. For example, there is reduced reproductive fitness and production of developmentally abnormal offspring in a zone of hybridization between two populations of velvet worms in eastern Australia. A less severe case of outbreeding depression is found in corroboree frogs in the mountains of southeastern Australia. Allopatric populations in the north and south of the range have subtle differences in colour pattern and skin alkaloids and exhibit a genetic distance of 0.17–0.49. Hybrids between the populations are fertile and viable, but display 17% larval abnormality, compared to <4% in crosses within regions. Populations differentiated due to genetic drift, rather than adaptation, are less likely to exhibit outbreeding depression. Thus, it is important, but difficult, to determine if population differentiation is due to drift or adaptation.

Tiger

Outbreeding depression may also occur when populations within a species that have adapted to different habitats are crossed

Extent of outbreeding depression in animals and plants

The extent and significance of outbreeding depression is a matter of controversy. Evidence in mammals and birds is scarce, apart from cases where the taxonomy has not been adequately resolved. The most widely quoted, but anecdotal, example is European ibex living in the Tatra Mountains of Czechoslovakia. The population was eliminated by adding desert-adapted animals of different sub-species from Turkey and from Sinai to European stock. The cause of extinction was disruption of the breeding cycle, with maladapted hybrids mating in early autumn and, fatally, giving birth in February, the coldest month.

Outbreeding depression has been detected for some traits in experimental crosses between different sub-species of mice, but this was small in comparison to the beneficial effects of crossing (heterosis). The endangered Arabian oryx is suffering simultaneously from both inbreeding depression and outbreeding depression, but there are chromosomal differences segregating in this population. Outbreeding depression has been clearly documented in an intertidal copepod that shows marked genetic differentiation, and limited dispersal, over relatively short geographic distances.

In several cases where outbreeding depression has been detected in animals, the presumed taxonomy proved to be erroneous. For example, crosses between different populations of owl monkeys and dik-diks resulted in sterile offspring. In both the cases the two contributing forms came from different localities, had different chromosome numbers, and were most probably undescribed species or sub-species.

Outbreeding depression is more frequent when crosses occur between populations that have undergone significant adaptation to local conditions and when dispersal is limited. Not surprisingly, most evidence for outbreeding depression comes from plants with these characteristics.

Outbreeding depression in crosses between populations within species has been documented in only a few cases in animals, but may be more common in plants

Even if crosses of populations result in outbreeding depression, natural selection will usually lead to rapid recovery and often higher eventual fitness

Currently, great caution is expressed about mixing populations. For example, doubts about the wisdom of crossing different populations of gray wolves have been expressed, even though they clearly belong to the same species. Many population geneticists consider that concerns over outbreeding depression, for species whose taxonomy is clearly understood, are often unjustified. Outbreeding depression is frequently mentioned, but the benefits of crossing are understated.

Even where outbreeding depression does occur when two partially inbred and differentiated populations are crossed, it will not be a long-term phenomenon. Unless the F_1 hybrid individuals are sterile, or of very low fitness, natural selection will act upon the extensive genetic variation in the hybrid population adapting it to its environment. The hybrid population will usually, at worst, go through a temporary decline in fitness, and then increase.

Defining management units within species

> Populations within species may be sufficiently differentiated in adaptive characteristics, or genetic composition, to require separate management

Species clearly require management as separate units. However, populations within species may be on the path to speciation. If they show significant adaptive differentiation to different habitats, or significant genetic differentiation, then they may justify management as separate evolutionary lineages for conservation purposes. The desirability of separate management depends on the balance between the cost of keeping two (or more) populations versus one, and risks of outbreeding depression, or benefits (higher genetic diversity and fitness) accruing from merging the populations.

Defining management units within species is more difficult and controversial than defining species. Below we outline the concept of evolutionarily significant units, together with a more recent proposal to define such units based on genetic and ecological exchangeability.

Evolutionarily significant units (ESU)

> An evolutionarily significant unit (ESU) is a population that has a high priority for separate conservation

Most conservation biologists believe that genetically differentiated populations within species should not be merged and require separate genetic management. These populations are referred to as **evolutionarily significant units**. Initially the concept was applied to populations with reproductive and historical isolation and adaptive distinctiveness from other populations within the species. Moritz proposed that genetic markers be used to define management units within species. If mtDNA genotypes show no overlap between populations, and nuclear loci show significant divergence of allele frequencies, then they should be defined as separate ESUs and managed separately. In broad terms this often means that well-defined sub-species are the unit of management, but such may not be the case in little-studied groups.

While many ESUs have been defined in threatened species, the concept has been criticized. In particular, ESUs defined solely using molecular genetic markers ignore adaptive differences. ESUs are unlikely to be detected within species with high gene flow, even though populations may have adaptive differences and warrant separate management. Conversely, in taxa with low gene flow, populations that have differentiated by genetic drift may be designated as separate ESUs, even though they may not be adaptively distinct – in this case they may benefit from gene flow.

Defining management units on the basis of exchangeability

Crandall and colleagues have recently proposed that management units within species be based upon whether populations are ecologically or genetically exchangeable, or replaceable by one another. This proposal attempts to delineate whether there is adaptive differentiation, whether there is gene flow, and whether differentiation is historical, or recent. If two populations are adapted to similar environments then they are exchangeable, but if they are adapted to different environments then they are not ecologically exchangeable. If the genetic compositions of two populations are similar, they are exchangeable, but if they are genetically differentiated they are not genetically exchangeable. The authors claim that this system deals more adequately with many cases where the ESU process yielded outcomes of doubtful justification.

> Management units can be defined using ecological and genetic exchangeability

In practice, the populations are given + (reject exchangeability) or − (accept) classifications in each of four cells, representing recent and historical genetic and ecological exchangeability (Fig. 6.3). This results in 16 categories of divergence between two populations. In general, the more + scores the greater the differentiation. Genetic exchangeability is concerned with the limits of spread of new genetic variation through gene flow. Exchangeability is rejected when there is evidence of restricted gene flow between populations, while it is accepted when there is evidence of ample gene flow. Evidence for gene flow is ideally based on multiple nuclear loci (allozymes, microsatellites, etc).

Ecological exchangeability is rejected (+) where there is evidence for population differentiation due to natural selection, or genetic drift. Evidence can be based on difference in life history traits, morphology, habitat, or loci under selection, and such differences should, ideally, be demonstrably heritable. We interpret this as primarily reflecting adaptive differentiation.

The recent and historical time frames are designed to distinguish natural evolutionary processes of limited gene flow from recent population isolation. Further, they distinguish secondary contact from long-term gene flow.

Recommended management actions are given for each of the eight categories in Fig. 6.3. Example 6.3 illustrates the application of the methodology to black rhinoceros and puritan tiger beetles. This

H_o exchangeability

Time frame

	Genetic	Ecological
Recent		
Historical		

Relative strength of evidence	Evidence of adaptive distinctiveness	Recommended management action
8	$\frac{+\,\mid\,+}{+\,\mid\,+}$	Treat as long-separated species
7	$\frac{+\,\mid\,+}{-\,\mid\,+}\quad\frac{+\,\mid\,+}{+\,\mid\,-}$	Treat as separate species
6	$\frac{-\,\mid\,+}{+\,\mid\,+}$	Treat as distinct populations (recent admixture, loss of genetic distinctiveness)
5	$\frac{+\,\mid\,-}{+\,\mid\,+}$	Natural convergence on ecological exchangeability – treat as single population.
4	(a) $\frac{+\,\mid\,+}{-\,\mid\,-}$ (b) $\frac{-\,\mid\,-}{+\,\mid\,+}$ (c) $\frac{-\,\mid\,+}{-\,\mid\,+}$	Anthropogenic convergence – treat as distinct populations. (a) & (b) Recent ecological distinction, so treat as distinct populations. (c) Allow gene flow consistent with current population structure
3	$\frac{-\,\mid\,+}{-\,\mid\,-}$	Allow gene flow consistent with current population structure; treat as distinct populations
2	$\frac{+\,\mid\,-}{+\,\mid\,-}$	Allow gene flow consistent with current population structure; treat as a single population
1	$\frac{+\,\mid\,-}{-\,\mid\,-}\quad\frac{-\,\mid\,+}{-\,\mid\,+}\quad\frac{-\,\mid\,-}{+\,\mid\,-}$ $\frac{+\,\mid\,-}{-\,\mid\,+}\quad\frac{-\,\mid\,+}{+\,\mid\,-}\quad\frac{-\,\mid\,-}{-\,\mid\,-}$	Treat as single populations; if exchangeability is due to anthropogenic effects, restore to historical condition; if natural, allow gene flow

Fig. 6.3 Defining management units within species on the basis of genetic and ecological exchangeability (Crandall et al. 2000). This method has been proposed for measuring categories of population distinctiveness, and consequent management actions recommended for each of the categories. Categories of population distinctiveness are based on rejection (+) or failure to reject (−) the null hypothesis of genetic and ecological exchangeability for both recent and historical time frames. As the number representing relative strength of the evidence increases, so does the evidence for significant population differentiation.

methodology provides a logical means for delineating populations that justify separate management, without having an excessive number of management units that do not show adaptive differentiation. By contrast, the term ESU has been applied to every category in Fig. 6.3, leading to a very large number of management units, some with doubtful justification.

| **Example 6.3** | Assignment of black rhinoceros and puritan tiger beetles to the exchangeability categories |

Black rhinoceros

BLACK RHINOCEROS
For the black rhinoceros in Africa, there are insufficient grounds to reject either genetic exchangeability or ecological exchangeability. Populations show gene flow and their habitats are similar. Consequently, it is categorized as case 1 ($- - / - -$), leading to the recommendation that the species be managed as a single population. Conversely, mtDNA data have been used to argue for two sub-species with separate management.

PURITAN TIGER BEETLE
Puritan tiger beetles from Connecticut River and Chesapeake Bay, USA are not genetically exchangeable, based on mtDNA (low gene flow and significant differentiation). Further they are not ecologically exchangeable based on habitat parameters. Thus they were classified as category 7 ($+ + / + -$), indicating strong adaptive differentiation. The recommendation was to manage the two populations as separate units for conservation purposes. Populations on the east and west of Chesapeake Bay were genetically and ecologically exchangeable ($- - / - -$).

SUGGESTED FURTHER READING

Frankham, R., J. D. Ballou & D. A. Briscoe. 2002. *Introduction to Conservation Genetics*. Cambridge University Press, Cambridge, UK. Chapter 15 has a slightly extended treatment of these topics, along with references.

Avise, J. C. & J. Hamrick. 1996. *Conservation Genetics: Case Histories from Nature*. Chapman & Hall, New York. Advanced scientific reviews that describe many cases of resolving taxonomic uncertainties with the aid of molecular genetic techniques.

Claridge, M. F., H. A. Dawah & M. R. Wilson. 1997. *Species: The Units of Biodiversity*. Chapman & Hall, London. A collection of scientific papers on species, including the many different definitions.

Crandall, K. A., O. R. P. Bininda-Edmonds, G. M. Mace & R. K. Wayne. 2000. Considering evolutionary processes in conservation biology: an alternative to 'evolutionarily significant units'. *Trends in Ecology and Evolution* 15, 290–295. Method proposed for using genetic and ecological differences between populations within species as a basis for deciding on whether they deserve separate management.

Futuyma, D. J. 1998. *Evolutionary Biology*, 3rd edn. Sinauer, Sunderland, MA. A textbook with an excellent readable coverage of speciation and the genetic processes underlying it.

Hall, B. G. 2001. *Phylogenetics Made Easy: A How-to Manual for Molecular Biologists*. Sinauer, Sunderland, MA. A clearly written guide to building phylogenetic trees from molecular data.

Nei, M. & S. Kumar. 2000. *Molecular Evolution and Phylogenetics*. Oxford University Press, New York. Textbook on genetic distances and molecular methods in taxonomy.

Chapter 7

Genetic management of endangered species in the wild

Genetic management of endangered species *in situ* involves recovery of small inbred populations, management of fragmented populations, alleviating genetic 'swamping' due to hybridization with related species, and minimizing the deleterious impacts of harvesting. Population viability analysis can be used to quantify threats and to compare alternative management options

Terms
Clones, corridor, introgression, metapopulation, sensitivity analysis, supplemental breeding, supportive breeding, translocation

The endangered red-cockaded woodpecker from the southeastern USA

Genetic issues in endangered populations

There has been only limited application of genetics in the practical management of threatened species in their natural habitats. Genetic issues are important in wild populations, but unfortunately are rarely considered, due largely to a lack of appreciation of their importance. We have already referred to some of the few examples and a number are presented in more detail here.

To summarize what we have discussed previously, the key genetic contributions to conservation biology are:

- resolving of taxonomic uncertainties so that managers can be confident of the status of, and relationships among, the populations they strive to maintain (Chapter 6)
- delineating of any distinct management units within species, as biologically meaningful entities for conservation (Chapter 6)
- detecting declines in genetic diversity. Sensitive genetic markers, particularly microsatellites, have the power to detect reductions in heterozygosity and allelic diversity in small and fragmented populations
- developing theory to describe past, and predict future, changes in genetic variation. All such theories have a common, central, element – genetic diversity is dependent on N_e
- recognizing that the effective sizes of populations are frequently about an order of magnitude lower than census sizes
- recognizing that loss of genetic diversity underlying quantitative variation in reproductive fitness reduces the capacity of populations to evolve in response to environmental change
- detecting of inbreeding depression in endangered species in natural habitats
- recognizing that inbreeding depression is an expected result of extensive inbreeding in almost all cases
- recognizing that potential inbreeding depression may be inferred from its correlation with reduction in genetic variation
- recognizing that degree of fragmentation and rates of gene flow can be inferred from the distribution of genetic markers within and among populations.

Several of these issues are illustrated by the example of the Florida panther (Box 7.1).

Box 7.1	Identifying genetic problems in the Florida panther and genetic management to alleviate them (Roelke *et al.* 1993; Barone *et al.* 1994; Land & Lacy 2000)

The endangered Florida panther is restricted to a small relict population of approximately 60–70 individuals in southern Florida, primarily in the Big Cypress and adjoining Everglades National Park ecosystems. Prior to European settlement, they ranged across the entire southeast of the USA, and other sub-species were spread throughout North and South America. Since 1973, the main causes of deaths have been road kills, illegal hunting or injuries. A population viability analysis in 1989 (see Chapter 5) predicted that this population had a high probability of extinction within a short time, unless remedial actions were taken. A more recent assessment is less pessimistic.

Analyses using allozymes, morphology and mtDNA revealed that a portion of the population had previously received genetic input (introgression) from a South America puma sub-species. Subsequently, records revealed that South American animals had been released into the population by a private breeder between 1956 and 1966. Hybrid animals are located in areas away from most 'authentic' animals.

The authentic population has very low levels of genetic diversity compared to the hybrids, other puma populations and felids generally.

Florida panther

Sub-species	Allozymes heterozygosity % (range)	DNA fingerprint heterozygosity %
Florida (authentic)	1.8	10.4
Florida (introgressed)	1.8	29.7
Western US pumas	4.3 (2.0–6.7)	46.9
Other felids	3.0–8.0	–
Domestic cat	–	44.0

Authentic Florida panthers also display evidence of inbreeding depression, including morphological abnormalities ('cowlick' patterns in their fur and kinked tails), cardiac defects and the poorest semen quality of any felid. All 'pure' males have at least one undescended testis (cryptorchidism). Florida panthers also suffer a high prevalence of infectious disease.

Normal   Coiled tail, abnormal acrosome

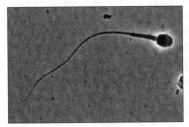
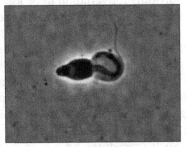

How can the authentic Florida panther be recovered? Since it is an extremely small population, the first priority is to increase population size through protecting and improving additional habitat and decreasing existing threats. Construction of culverts under highways significantly reduced highway mortality. Since these panthers display both the genetic and physical hallmarks of inbreeding depression, their fitness was increased by immigration from another population. As no other population exists in Florida, the only source of individuals was thought to be from another sub-species. The nearest sub-species is from Texas, a population that was contiguous with the Florida population and probably had historical gene exchange before its decline.

Following extensive consultations, the decision was made to introduce six to eight Texas females. Outbreeding depression was unlikely, as there was no evidence for it, either in the hybrid Florida animals or in mixed populations of captive pumas. While ideally genetic analyses comparing the sub-species should have been conducted prior to making the decision, subsequent molecular analyses showed that all North American panthers/cougars are very similar.

Eight individuals from Texas have now been introduced. Thirty-two surviving outcrossed progeny are known (as of June 2001), including some second generation and backcross offspring from hybrids. The F_1 hybrid kittens lack cowlicks and kinked tails and appear to be more robust than authentic Florida panthers.

This chapter follows approximately the order of genetic management actions for wild populations, as shown in the flow chart in Fig. 7.1. As in all conservation, before we can devise a management plan we must know what we are conserving. For the rest of this chapter, we will assume that taxonomic uncertainties and management units have been resolved (Chapter 6), and address genetic management within units.

Increasing population size

A primary objective in managing wild populations is to increase the sizes of small populations

A primary objective in threatened species management is to reverse population decline. This simultaneously alleviates most of the stochastic (demographic, environmental, catastrophic and genetic) threats. If populations have recently declined from large sizes to $N_e \geq 50$ and are rapidly expanded, then the genetic impacts are minimal. Despite the bottleneck, short-term reductions of this magnitude allow little opportunity for variation to be lost through genetic drift (Chapter 4).

The first step in this process is identifying and removing the causes of the decline. This is the domain of wildlife biologists and ecologists. Actions taken include legislative controls on hunting and harvesting, designation of reserves, reduction of pollutants, improvement of habitat quality and eradication of unnatural predators and competitors. These actions typically benefit many native species in the managed region. In this respect, the endangered species acts as a 'flagship' for an entire community.

Such procedures have been successful in a diversity of species. For example, Indian rhinoceroses have increased from 27 to about 600 in Chitwan National Park, Nepal following bans on hunting and designation of the Royal hunting reserve as a national park. Similarly, the northern elephant seal has expanded from 20–30 to over 150 000 following cessation of hunting and legislative protection. Mauritius kestrel, bald eagle and peregrine falcon populations have all recovered following control of DDT usage, typically in combination with assisted breeding.

Indian rhinoceros

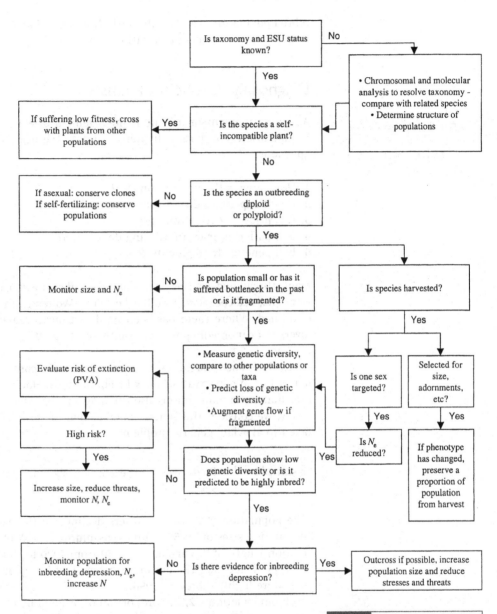

Fig. 7.1 Flowchart of questions asked in the genetic management of threatened species in the wild.

The Chatham Island black robin has increased from five to over 140 following protection, cross-fostering and translocations. The Lord Howe Island woodhen population has recovered from 20–30 to around 200 following eradication of their primary threat, pigs, and a short-term captive breeding and release program. In plants, recovery has followed legislative protection to minimize harvest, creation of reserves and removal of exotic herbivores. Where numbers are extremely small, forms of *ex situ* conservation have been used, followed by reintroductions (Chapter 8).

While genetic information may help to alert conservation biologists to the extent of endangerment, the management actions above involve little, or no, genetics. However, recovery in numbers in highly inbred populations can be substantially enhanced following introduction of additional genetic variation (see below).

Diagnosing genetic problems

Much of the contribution of genetics to management of wild populations has been diagnosis of genetic status

A necessary precursor to genetic management of wild populations is to diagnose their status (Chapters 2 and 4). We need to answer five questions:

1. How large is the population (N_e)?
2. Has it experienced significant bottlenecks in the past?
3. Has it lost genetic diversity?
4. Is it suffering from inbreeding depression?
5. Is it genetically fragmented?

Many threatened species including cheetahs, northern hairy-nosed wombats, red-cockaded woodpeckers and Wollemi pines have been examined. Where there has been no direct measurement of genetic diversity or inbreeding, we can use theory to predict them, as shown in Example 7.1.

To date, such diagnoses have been the main contribution of genetics to the conservation of wild populations. However, using this information to plan conservation management is still in its infancy. Below, we consider the genetic management actions that should be taken to alleviate genetic problems.

Example 7.1 | Predicting loss of genetic diversity and inbreeding due to small population sizes

The population of Florida panthers declined in the early 1900s to population sizes of 30–50 animals. Assuming a N_e/N of 0.10, a generation length of 7 years, and $N = 50$ from 1920 to 1990, would we have predicted that the Florida panthers had low genetic diversity and were highly inbred in 1990?

From equation 4.2, the proportion of original heterozygosity remaining in the population is:

$$\frac{H_t}{H_0} = \left(1 - \frac{1}{2N_e}\right)^t = \left(1 - \frac{1}{2 \times 5}\right)^{10} = 0.9^{10} = 0.35$$

as $N_e = 50/10 = 5$ and 70 years = 10 generations.

Based on theory alone, we would have predicted that about 65% of its heterozygosity had been lost, a value roughly in agreement with the observed loss inferred from comparing Florida panthers with pumas in the Western US (Box 7.1). From equation 4.4, the inbreeding coefficient is predicted to be 0.65, a level of inbreeding where deleterious effects would clearly be expected.

Recovering small inbred populations with low genetic diversity

An effective management strategy for recovering small inbred populations with low genetic diversity is to introduce individuals from other populations to improve their reproductive fitness and restore genetic diversity. There is extensive experimental evidence that this approach can be successful. For example, increases in fitness and population sizes have occurred following introductions of immigrants into a small isolated population of adders in Sweden, and into the small Illinois population of greater prairie chickens (Fig. 7.2).

Despite the clear benefits of this procedure, there are very few cases where it is being practiced (see below).

> Small, inbred populations can be recovered by introducing unrelated individuals

Swedish adder

Source of unrelated individuals for genetic augmentation

The individuals chosen for introduction into inbred populations for recovery of fitness and genetic diversity, may be either

- outbred (if available), or
- inbred but genetically differentiated from the population to which they are being introduced.

An example of the latter situation is provided by Australian wallabies (Chapter 5). The black-footed rock wallaby has several inbred island populations that could be combined to increase genetic diversity and improve reproductive fitness. The combined genetic material from all the island populations would contain most of the microsatellite alleles found within the mainland populations. In the future, it may be necessary to use crosses among island populations to reconstitute a

> Individuals from genetically distantly related populations, or from related inter-fertile taxa, can be used to augment small inbred populations

Greater prairie chicken

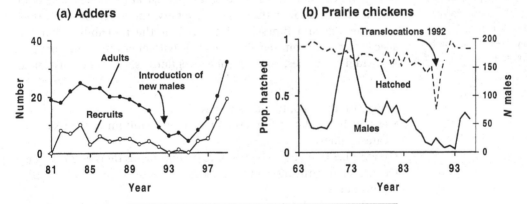

Fig. 7.2 Recovery of reproductive fitness due to introduction of immigrants into small partially inbred populations of (a) adders in Sweden (after Madsen *et al.* 1999), and (b) greater prairie chickens in Illinois – translocations began in 1992 (Westemeier *et al.* 1998).

new population for reintroduction to mainland localities once foxes, their major threat, are eliminated.

Where no unrelated individuals of the same taxon are available, individuals from another sub-species can be used to alleviate inbreeding depression, for example in the Florida panther (Box 7.1) and the Norfolk Island boobook owl (Chapter 6).

If an endangered species persists only as a single population, then the only possible source of additional genetic material is from a related species to which they can be successfully crossed. A blight from China severely depleted the American species of chestnut. The Chinese chestnut species is resistant to the blight and has been crossed with the American species to introduce resistance alleles.

The option of crossing a threatened species to a related species requires extreme caution and must be evaluated on an experimental basis, to ensure that hybrids are fertile and viable in F_1 and subsequent generations, prior to full implementation. The potential benefits must be very large, as there may be a serious risk of outbreeding depression. Even where some outbreeding depression occurs, inbreeding depression will be greatly reduced and natural selection will eventually remove most or all of the outbreeding depression. Given that the extent of genetic differentiation among species varies considerably, crosses among species in some taxa will only be equivalent to crosses between sub-species in other taxa (Table 6.1).

Management of species with a single population lacking genetic diversity

For species consisting of a single population with reduced genetic diversity, the only options are to improve their environment and minimize risks associated with changed environments (especially disease) and small population size

From a genetic perspective, the worst situation is an endangered species with only a single inbred population, with no sub-species or related species available for augmentation. Information on loss of genetic diversity is useful only as an indicator of the fragility of the species. The lower the genetic diversity, the lower the evolutionary potential and fitness, and the higher the probability that the species has compromised ability to cope with environmental changes. For fragile species, management regimes should be instituted to:

* increase their population size (see above)
* establish populations in several locations (to minimize the risk from catastrophes)
* maximize their reproductive rate by improving their environment
* consider instituting captive breeding or other *ex situ* conservation procedures
* insulate them from environmental change.

The latter regime should include quarantining from introduced diseases, pests, predators and competitors and monitoring, so that remedial action can be initiated as soon as new environmental threats arise. For example, the recently discovered and endangered Wollemi pine in Australia has no genetic diversity within or among populations (Chapter 1). The recovery plan for this species calls for (a) restricting access to the populations by keeping their location secret, (b) limiting access to approved people, (c) instituting strict hygiene protocols to avoid introducing disease, (d) fire management and (e) maintaining *ex situ* samples of each plant in botanic gardens. Further, commercial propagation of the species is increasing population size and reducing stochastic risks.

The black-footed ferret has low genetic diversity, but is a less extreme case. Its recovery plan calls for creation of a captive population to re-establish 10 wild populations in different locations to minimize the risks of disease and other environmental catastrophes.

Genetic management of fragmented populations

Many threatened species have fragmented habitats, of the type illustrated for the giant panda (Fig. 7.3). The management options for fragmented populations to maximize genetic diversity and minimize inbreeding and extinction risk are to

> The adverse genetic consequences of population fragmentation can be alleviated by re-establishing historic levels of gene flow among fragments, by improving habitat quality and by re-establishing populations in areas where they have become extinct

Panda

Fig. 7.3 Habitat fragmentation for the giant panda in China (after Lu *et al.* 2001).

- increase the habitat area and quality
- artificially increase the rate of gene flow by translocation to match historic levels
- create habitat corridors, and
- re-establish populations in suitable habitat where they have gone extinct

> Gene flow must be re-established to reduce the risk of extinction in genetically isolated fragments

To alleviate or prevent deleterious genetic consequences in isolated fragments, gene flow needs to be re-established by moving individuals (**translocation**), or gametes, or by establishing ribbons of habitat between fragments as migration **corridors**. The benefits of immigration have been established in many cases. In the small scabious plant, between-population crosses had fitnesses 2.5 times that of within-population crosses. Computer projections indicate that immigration will reduce the extinction risks for two small black rhinoceros populations in East Africa (Box 7.2).

Translocation of individuals among populations may be costly, especially for large animals, and carries the risks of injury, disease transmission and behavioural disruption when individuals are released. For example, introduced male lions regularly kill cubs. Further, sexually mature males of many species may kill intruders. The cost of translocations can be reduced by artificial insemination for species where this technique has been perfected (see below). The same care to avoid outbreeding depression, discussed previously, must be exercised in planning translocations.

Corridors among habitat fragments (frequently recommended for non-genetic reasons) can re-establish gene flow among isolated populations. Species vary in their requirements for a corridor to be an effective migration path. The most ambitious proposal of this kind is 'The Wildlands Project' – to provide corridors from north to south in North America. The corridors will link existing reserves and both reserves and corridors will be surrounded by buffer zones that are hospitable to wild animals and plants. The time frame for achieving this vision is hundreds of years, given the political, social and financial challenges. Nonetheless, such systems are essential if we are to conserve biodiversity in the long term. With global climate change, plants and animals need to alter their distributions to cope with moving climatic zones. At present, such movement is often prevented by inhospitable habitat between protected areas

| **Box 7.2** | Modelling the effects of inbreeding depression and immigration on the survival of black rhinoceros populations in Kenya (after Dobson et al. 1992) |

Black rhinoceros populations are threatened across their entire range, due predominantly to poaching. In East Africa, no single population numbered more than 60 animals in the early 1990s and the situation is probably worse now. The

Black rhinoceros

distribution is fragmented and there is no gene flow among most, or all, of the fragments.

Dobson and co-workers investigated the demographic and genetic factors likely to contribute to extinction risk and the factors that must be considered in management. Stochastic computer simulations (PVA) were used to project the history of each individual from birth to death, including their reproductive output (Chapter 5). Input files, using data from sanctuary populations, contained the identity and parentage of each individual, age and sex-specific rates of survival and reproduction together with rates of immigration of different age and sex classes.

Populations at Lewa Downs and Nakuru National Park, Kenya were modelled for 200 years with (a) no immigration and no inbreeding depression (only demographic and environmental stochasticity), (b) with the addition of inbreeding depression, but with no immigration and (c) with inbreeding depression and immigration of one immigrant every 10 years for the next 50 years. Inbreeding depression on survival was applied at a level approximating the average value observed for captive mammal populations. The initial population sizes of the two populations were set at their actual sizes, 10 females and 3 males at Lewa Downs and 7 females and 11 males at Nakuru National Park.

In case (a), stochastic factors led to an extinction risk of over 50% after 200 years in the smaller Lewa Downs population, but less than 10% in the larger Nakuru population, both due to stochastic factors. With inbreeding depression (case b), the extinction probabilities rose to 76% and 22%, respectively. With inbreeding depression and immigration, extinction risks were lowest being about 40% and 5%, respectively (case c). Consequently, immigration is predicted to reduce extinction risk in these fragmented rhinoceros populations. Additional migration after year 50 would result in further reductions in extinction risks.

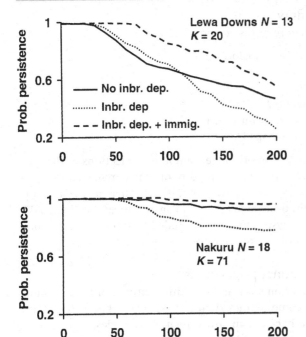

Despite the widespread fragmentation of populations of threatened species, we are aware of few cases in animals and plants where immigration is being used as a practical measure to alleviate inbreeding and loss of genetic diversity in the wild. Individuals from other populations are being introduced to small populations of endangered red-cockaded woodpeckers, as computer simulations have predicted their likely extinction without augmentation. Other animal cases are the endangered and inbred Florida panther (Box 7.1) and the golden lion tamarin (Box 8.3).

Two botanical examples involve self-incompatible plants where loss of alleles at self-incompatibility loci has reduced fitness due to shortage of compatible pollen (Chapter 5). The threatened lakeside daisy population in Illinois was incapable of reproducing as it contained so few self-incompatibility (S) alleles. It was outcrossed to Ohio plants and seeds from Ohio and Ontario populations were introduced to increase the diversity of S alleles and recover reproductive fitness.

Over-grazing by sheep reduced the Mauna Kea silversword to 50 adult plants. Subsequently, 450 plants were raised and outplanted to augment the populations. However, genetic analyses revealed that all the outplanted individuals were the progeny of only two female plants and had lower genetic diversity than the original wild population. As this is a self-incompatible species, the outplanted individuals produced only a 20% seed set. To remedy these problems, outplanted individuals have been cross-pollinated using pollen from unrelated individuals from the natural population, resulting in a 60% seed set presumably because this pollen carries different S alleles.

Mauna Kea silversword

Managing gene flow

Managing gene flow requires consideration of:

- which individuals to translocate?
- how many?
- how often?
- from where to where?
- when should translocation begin?
- when should it be stopped?

With so many variables to optimize, computer projections of the type given in Box 7.2 will often be required to optimize management. The objective is to identify a regime that maintains genetically viable populations, fits within other management constraints and is financially feasible. Genetic analysis should be included in monitoring the program to allow application of adaptive management policies.

Re-establishing extinct populations

To maximize population sizes and minimize extinction risks, populations that have become extinct should be re-established from extant populations, if the habitat can still support the species.

Which populations should be used to re-establish extinct populations? To minimize inbreeding and maximize genetic diversity, the re-founding population should be sampled from most, or all, extant populations. A case of poor choice of populations is provided by the koala in southeastern Australia (Box 7.3). Island populations with low genetic diversity were used for reintroduction, as ample individuals were available from those sources. Genetic issues were ignored and deleterious genetic consequences have resulted.

Where there is evidence of adaptive genetic differentiation among extant populations (e.g. in many plants), the translocated individuals should normally come from populations most likely to be the best adapted to the reintroduction habitat. This will frequently be the geographically closest surviving population.

Re-establishment should be done using the most genetically diverse population with the highest reproductive fitness in the release environment, or a cross among populations, provided all alternative populations have similar adaptation to the release site.

| Box 7.3 | Reintroduction of koalas in southeastern Australia: a poorly designed program with adverse genetic impacts (Houlden et al. 1996; Sherwin et al. 2000; Seymour et al. 2001) |

Koala

The koala is a unique marsupial endemic to eastern Australia. It is both a cultural icon and an important contributor to tourist income. It once ranged from Queensland to Victoria and South Australia, but its numbers have been reduced by hunting, habitat loss and disease. By the 1930s, koalas inhabited less than 50% of their former range, had disappeared in South Australia, and were nearly extinct in Victoria. However, they were still considered common in Queensland where they recovered without large-scale assistance. The fur trade ceased by the 1930s when koalas

were given legal protection in all States. Subsequently, much effort has gone into koala conservation.

Extensive translocations of animals occurred in the southeast. A population was founded from as few as two or three individuals on French Island (FI) in Victoria late in the nineteenth century. In the absence of predators, this population rapidly reached carrying capacity. Surplus animals from French Island were used to found an additional population on Kangaroo Island (KI) (18 adult founders plus young) in 1923–25, and to supplement a population founded on Phillip Island (PI) in the 1870s. The French and Phillip Island populations were used widely to supplement mainland Victorian populations and the Kangaroo Island population to restock mainland South Australia. The restocked South Australia mainland population has gone through three bottlenecks, mainland Victoria → French Island → Kangaroo Island → mainland South Australia. Since 1923, 10 000 animals have been translocated to 70 locations.

Stocking of populations using individuals from bottlenecked populations resulted in loss of genetic diversity and inbreeding. The populations in Victoria and South Australia have about half the genetic diversity of the less-perturbed populations further north. The Kangaroo Island population has the lowest genetic diversity of all surveyed populations. All southeastern populations show similar microsatellite allele frequencies, while the more northerly populations exhibit considerable genetic differentiation among populations. The translocations have also led to an increased frequency of males with missing testicles (testicular aplasia), this being worst in the most bottlenecked population on the South Australian mainland.

What might have been done to avoid such problems? Loss of genetic diversity and inbreeding depression would have been averted if the French Island population had been founded with more individuals, or if its genetic diversity had been augmented to give it a greater base.

What can be done to reverse the current problems? The most efficient strategy would be to introduce more genetic diversity into the southeastern populations (both island and mainland) from nearby populations with high genetic diversity.

Genetic issues in reserve design

Reserves need to be sufficiently large to support the target species, and contain habitat to which the species is adapted. Natural or artificial gene flow must occur between reserves

Many ecological, political and genetic issues must be balanced in designing nature reserves. Soulé & Simberloff suggested that three steps should be involved: (a) identify target or keystone species whose loss would significantly decrease the biodiversity in the reserve, (b) determine the minimum population size needed to guarantee a high probability of long-term survival for these species, and (c) using known population densities for these species, estimate the area required to sustain minimum numbers. The genetic issues in reserve design are:

- is the reserve large enough to support a genetically viable population?
- is the species adapted to the habitat in the reserve?
- should there be one large reserve, or several smaller reserves?

From the arguments given in Chapter 5, a reserve of sufficient size to maintain an effective population size of at least several hundred and an actual size of several thousand is required.

Should threatened species be maintained in a single large reserve, or in several smaller reserves? In general, a single large reserve is more desirable from the genetic point of view, if, as is probable, there is a risk that populations in small reserves will become extinct. In contrast, protection against catastrophes dictates that more than one reserve is required. The best compromise is to have more than one sizeable reserve, but to ensure that there is natural, or artificial, gene flow among them. In practice, the choice has often been a haphazard process, determined more by local politics, alternative land uses and the need for reserves to serve multiple purposes (e.g. recreation), than by biological principles.

Introgression and hybridization

Introgression is the flow of alleles from one species or sub-species to another. Typically, hybridizations occur when humans introduce exotic species into the range of related rare species, or alter habitat so that previously isolated species are now in secondary contact. Introgression is a threat to the genetic integrity of a range of canid, duck, fish, plant, and other species.

Introgression can be detected using a wide range of genetic markers, including allozymes, microsatellites, DNA fingerprints and, where appropriate, chromosomes. Allozyme analyses of captive 'Asiatic' lions detected alleles from African lions. Endangered Florida panthers consist of two populations, one of which resulted from introgression from a South American sub-species of panthers (Box 7.1). Females of the highly endangered Ethiopian wolf have been shown to hybridize with male domestic dogs, with the levels varying in different populations (Example 3.1).

A particularly deleterious form of hybridization is that between related diploid and tetraploid populations, resulting in sterile triploids, for example in the endangered grassland daisy in Australia.

Alleviating introgression

The Catalina mahogany (Box 7.4) provides an example of management of a species subject to hybridization. In this case, genetic information was critical in defining the hybridization problem and in identifying hybrid individuals, but had a limited role in the recovery process.

Options for addressing the problem of introgression include eliminating the introduced species, or translocating 'pure' individuals into isolated regions or into captivity. Success is hard to achieve. For example, it will not be practical to remove all domestic dogs from the habitat of the Ethiopian wolf.

> Some species are endangered due to genetic 'swamping' by interbreeding with abundant related species. Genetic analyses can be used to detect introgression and hybrid individuals

Endangered grassland daisy

> Introgression can be reduced by removing the hybridizing species, by eliminating hybrid individuals, or by expanding the number of 'pure' individuals

Catalina mahogany

Impacts of harvesting

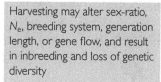

Harvesting may alter sex-ratio, N_e, breeding system, generation length, or gene flow, and result in inbreeding and loss of genetic diversity

Many species of wild animals and plants are harvested or poached, e.g. fish, trees, deer and elephants. This may reduce effective population size and genetic diversity with deleterious consequences. For example, poaching has impacted on sex-ratio, reproductive rate and effective population size in Asian elephants (Box 7.5). Further, hunting of moose and white-tailed deer has been projected to severely reduce genetic diversity.

Asian elephant

Selective harvesting by humans may change the genetic composition and phenotypes of populations

Selective harvesting may favour particular phenotypes within populations e.g. large fish, tusked elephants, and deer with large antlers are taken. This may result in selection pressures that change the phenotype of the species, conflicting with forces of natural selection and reducing the overall fitness of the population. For example, ivory poaching is increasing the frequency of tuskless male elephants in several populations in Africa and Asia. Tuskless elephant males may be less successful than tusked males in finding mates and less able to ward off predators.

The obvious solution to this problem is to reduce the selectivity of the harvest, but this may be very difficult to achieve in practice. Harvest of elephant ivory is itself an illegal activity, not subject to regulation. Despite many international agreements, total fish catches are difficult to regulate, as the collapse of many fisheries attests. As harvested species often occur in large, though frequently declining numbers, an option is to preserve a proportion of the population from harvest. In this way, fully wild stocks are maintained to repopulate harvested areas and minimize the genetic impacts of harvest.

The genetic and evolutionary impacts of selective poaching have received very little consideration in the management of endangered species.

> Impacts of selective harvest may be alleviated by changing harvest regimes, or by preserving a portion of the species without harvest

Genetic management of species that are not outbreeding diploids

Many species of plants and some animals are not diploid outbreeders and so require modified management regimes. These include:

- asexually reproducing species
- self-fertilizing species
- polyploid species (found in many plants and a few animals).

Below we consider their characteristics and the modifications to genetic management that these entail.

> Species that are not diploid outbreeders require modified genetic management as they differ in their distributions and rates of loss of genetic diversity and in their responses to inbreeding

Asexual species

Many species of plants are capable of vegetative reproduction, via runners, bulbs, corms, etc. Species with exclusively asexual reproduction typically exist as one, or a few, **clones**. Genetic diversity may occur among clones, but individuals within clones are essentially identical. For example, the endangered Australian triploid shrub King's lomatia exists as a single clone and the endangered *Limonium dufourii* plant from Spain consists of several triploid clones.

> In asexual species, many individuals may have identical genotypes, so that great care is required to sample the diversity of clones within the species

In a fully clonal species, inbreeding is not a concern. However, species that exist as a single clone will have essentially no variation for adapting to environmental changes. They parallel inbred populations of outbreeding species. This fragility requires management of the type used for the Wollemi pine (see above).

In species that employ both sexual and asexual reproduction, the number of genetically distinct individuals may be far less than the number of individuals. For example, 53 individuals of endangered Australian shrub *Haloragodendron lucasii* consisted of only seven clones, with reduced genetic diversity, compared to an equivalent sexually reproducing species.

Maintenance of genetic diversity in asexually reproducing species requires that the structure of such populations be recognized for *in situ* conservation, in re-establishing extinct populations, and in sampling for *ex situ* conservation. The major conservation need is to identify and maintain as many distinct clones as possible.

Self-fertilizing species

About 20% of plant species habitually self-fertilize, e.g. the endangered Malheur wirelettuce, and another 40% do so occasionally, e.g. Brown's banksia.

> Selfing species typically have less heterozygosity within populations and more differentiation among populations than outbreeders. Consequently, much greater effort is required to preserve diverse populations than for outbreeders

Predominantly self-fertilizing species are typically less heterozygous than outbreeding species, with a higher proportion of their genetic diversity being distributed among populations than within populations. Populations frequently contain unique alleles. Consequently, alleles are more likely to be eliminated by the loss of individual populations in an inbreeding species than in an outbreeding species.

Inbreeding is less of an issue in selfing species, as they typically suffer less inbreeding depression than outbreeding species. However, many selfing populations outcross periodically and the opportunity to do so should be maintained in threatened species.

Reduced gene flow because of population fragmentation is less important in naturally inbreeding species, as they are already highly fragmented. To retain genetic diversity within populations, and heterozygosity within individuals, small fragmented populations need to be augmented with individuals from other fragments.

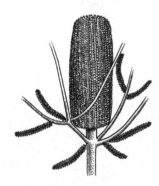

Brown's banksia

Polyploid species

> Genetic management issues for polyploid species are similar to those for diploid species, but there are fewer concerns about small population size and inbreeding

Genetic management of outbreeding polyploid species (the majority of angiosperms and ferns) follows the same principles as for diploids with similar breeding systems. However, genetic concerns are generally less serious than for equivalent diploids. For example, a population of tetraploids contains twice as many copies of each allele as a diploid population of the same effective size. Thus, polyploids probably suffer less inbreeding depression and loss of genetic diversity in small populations than diploids. Consequently, polyploids should tolerate lower population sizes than diploids. However, the sizes required to avoid demographic and environmental stochasticity and catastrophes will be similar to that for diploids (see Chapter 5).

Evaluating recovery strategies

> Population viability analysis is widely used as a management tool to compare options to recover species

Many procedures have been recommended for recovery of threatened populations, including legal restraints on exploitation, removing predators, improving habitat, reserving habitat, captive breeding, etc. Population viability analysis (PVA) is frequently used to evaluate and compare these options (Chapter 5). Typically, these commence with **sensitivity analyses**, followed by detailed PVAs that compare a range of specific management options.

Sensitivity analyses

Sensitivity analyses provide a valuable tool in assessing recovery programs for threatened species. These analyses involve varying different input parameters in PVAs by small amounts in either direction around the mean and evaluating their effects on extinction risk or population sizes. In this way, the parameter(s) whose values most influence the output can be identified. For example, is the outcome most sensitive to varying juvenile survival, or adult reproductive rate, or predation level, etc?

Alternative management options can then be compared. For example, first-year survival proved to be the parameter whose variation most impacted on population growth in the greater prairie chicken. Consequently, management should focus on improving nest success, brood survival and post-brood survival to one year of age.

Sensitivity analyses can alter our perceptions about the priority of threatening factors. For example, predation on cubs by lions and hyenas had been viewed as the major threat to the cheetah's viability. However, a sensitivity analysis revealed that this species was much more sensitive to changes in adult than juvenile survival. Similarly, many years of loggerhead turtle management focused on the seemingly obvious notion that increasing hatchling survival should reverse population decline. Sensitivity analysis demonstrated that reduction in adult mortality by capture in shrimp nets was the most efficient way to prevent decline.

> Sensitivity analyses involve determining the relative impact of different parameters on extinction risk

Case studies

Lord Howe Island Woodhen

The Lord Howe Island woodhen population declined to 20–30 individuals in the 1970s, through past human exploitation and, especially, from predation and habitat destruction by introduced pigs. The woodhen recovered following extermination of pigs and a captive breeding and reintroduction program where 86 captive-bred individuals were released over four years. A retrospective analysis of the recovery program confirmed that the management procedures implemented were indeed necessary. Extinction probabilities over 100 years were 100% for no management, 44% for pig control alone, and 2% for the combination of pig control and captive breeding. While pig control alone removed the main threatening process, it did not ensure that the population reached sufficient numbers to avoid extinction from stochastic factors.

Lord Howe Island woodhen

The woodhen now has a relatively stable population of about 200 on the island, and its status has been downgraded from endangered to vulnerable. Whilst the ecological side of this program is regarded as a model, the genetic management was far from optimal. As only three pairs contributed to the captive breeding program, the total population will be much more inbred than prior to the program. Further, genetic variation is likely to be lost due to the distorted contributions of the available individuals. In hindsight, it would have been desirable to enhance the genetic base of the captive breeding program once it was known to be successful, i.e. by adding further fertile individuals to the captive breeding program. Unfortunately, no genetic data on the population have been reported, no remedial action is planned and routine monitoring has ceased.

A prospective PVA concluded that the population is highly sensitive to minor changes in mortality and fecundity (e.g. due to inbreeding depression), and to catastrophes due to exotic species, or disease. The establishment of a second, remote population on another island was recommended to minimize these risks, but this has not been implemented.

Chinook salmon

Chinook salmon

Chinook salmon in Oregon, USA have declined dramatically since early last century, primarily due to habitat degradation associated with siltation from road building and forestry. The spring run in the South Umpqua River currently averages less than 300 spawners per year. A PVA on this population projected very low extinction risks over 100 and 200 years, assuming no further habitat degradation. However, this conclusion was highly sensitive to uncertainty about density dependence. The projected extinction risk was 100% assuming continuing habitat degradation at a rate similar to that in the past.

Furbish's lousewort

Furbish's lousewort

This endangered herbaceous perennial plant was once thought to be extinct. However, about 5000 individuals were discovered in 28 colonies along a 230-km stretch of a single river in Maine and New Brunswick in northeastern North America. The species is limited to periodically disturbed, north-facing riverbanks. It is an early successional hemiparasite that cannot invade disturbed riverbanks for at least three years, but is later crowded out by taller competitors, leading to regular rounds of colonization and population extinction (the species exists as a **metapopulation**). A PVA demonstrated that individual populations had 87% probabilities of extinction within 100 years, so that the survival of the species is critically dependent on the balance between colonization and extinction. As extinctions currently exceed colonization, the long-term viability of this species is tenuous. Further, the long-term ability of the population to adapt is questionable as four populations of the species lack genetic diversity at 22 allozyme loci.

Supplemental breeding and assisted reproductive technologies

Captive breeding is an important conservation option for species with a high risk of extinction in the wild. In the next chapter, we will consider genetic management of captive or *ex situ* populations. We therefore conclude this chapter by briefly outlining three issues which lie between wild and captive management.

Supplemental breeding

Supplemental breeding is widely used in fish management to augment wild populations. It differs from typical captive breeding in that wild adults are brought into captivity, bred and their progeny partially reared prior to release. There is no permanent captive population. The management of such programs has two goals that often conflict: the release of a large number of juveniles, and the maintenance of genetic diversity in the wild population. For highly prolific species, including many fish, only a small number of adults is required to produce the released juveniles, and the capture of only this small number may be desirable for endangered species. However, the procedure may reduce genetic diversity and increase inbreeding as the wild population experiences a reduced effective size through grossly unequal family sizes (Chapter 4). While the genetics of such programs initially received limited attention, it has received increasing focus in recent years.

> Supplemental breeding involves the capture of adults from nature, breeding in controlled settings and release of the offspring into the wild

Supportive breeding

Some wild populations that are not self-sustaining are maintained by regular augmentation from captive populations (**supportive breeding**). The nene (Hawaiian goose) has been subject to a long program of augmentation from captivity, as its wild population does not appear to be self-sustaining. Hatchery fish stocks are widely used to augment many wild fish species, especially those favoured by anglers.

> Some wild populations are regularly augmented from captive populations, but this is likely to be genetically deleterious in the long term

Such supportive breeding typically has deleterious long-term impacts on the genetic composition and reproductive fitness of the wild stock, namely: (a) reduced effective population sizes, (b) loss of genetic diversity, (c) inbreeding depression and (d) reduced reproductive fitness resulting from genetic adaptations to captivity that are deleterious in the wild. Genetic adaptation is a problem in fish when stocking of wild habitats is done using long-term captive populations, as they typically have lower reproductive fitness in wild habitats than do residents (see Chapter 8).

Nene

Assisted reproductive technologies

Assisted reproductive technologies, including artificial insemination, long-term gene banking, gamete rescue, cryopreservation of gametes, *in vitro* fertilization, embryo transfer, nuclear transfer (cloning) and seed and tissue banks, are rapidly advancing. The long-term aim is to supplement wild populations with captive-produced individuals, or even to resurrect extinct species from preserved tissue or genetic material. While much of this technology is developmental and is not available for most wildlife species, it has the potential to contribute significantly to conservation genetics in the following ways:

> Artificial insemination, long-term gene banking, gamete rescue, cryopreservation of gametes, *in vitro* fertilization, embryo transfer and nuclear transfer (cloning) can all contribute to conservation of threatened species

- transfer of genetic material between highly fragmented populations can be facilitated by cryopreserving semen from males in some fragments and artificially inseminating females in others
- gene banks for highly endangered species can preserve genetic diversity over the long term without the loss of alleles accompanying normal reproduction over many generations
- genetic contribution can be rescued from dead individuals by collecting their gametes shortly after death
- embryos from endangered species can be implanted into related species, increasing the reproductive output of the endangered species
- nuclear transfer and cloning can facilitate genetic management where numbers of individuals limit breeding opportunity.

The application of reproductive technology to genetic management of endangered animals has focused on managing captive populations, and has been used for only a few species. While these procedures have considerable potential for 'charismatic' endangered animal species where public concern supports the considerable expense involved, they are unlikely to be widely applicable. However, in plants, seed and tissue banks and cryopreservation have wide applications.

SUGGESTED FURTHER READING

Frankham, R., J. D. Ballou & D. A. Briscoe. 2002. *Introduction to Conservation Genetics*. Cambridge University Press, Cambridge, UK. Chapters 16 and 20 have extended treatments of these topics, along with references.

Avise, J. C. & J. L. Hamrick. (eds.) 1996. *Conservation Genetics: Case Histories from Nature*. Chapman & Hall, New York. Contains case histories on wild management of threatened species.

Beissinger, S. R. & D. R. McCullough. (eds.) 2002. *Population Viability Analysis*. University of Chicago Press, Chicago, IL. Proceedings of a conference on PVA that reviews the contribution of PVA to endangered species management.

Ecological Bulletin. 2000. Volume 48: *Population Viability Analysis*. A special volume devoted to the process of PVA and its contributions to conservation.

Falk, D. A. & K. E. Holsinger. 1991. *Genetics and Conservation of Rare Plants*. Oxford University Press, New York. Contains a number of chapters on wild management of threatened plant species.

Lanza, R. P., B. L. Draper & P. Damiani. 2000. Cloning Noah's ark. *Scientific American* 87 (5), 84–89. Describes actual and potential uses of artificial reproductive technologies in conservation.

Woodford, J. 2000. *The Wollemi Pine: The Incredible Discovery of a Living Fossil from the Age of the Dinosaurs*. Text Publishing, Melbourne, Australia. An interesting, well-written popular book on the discovery, conservation and genetics of the Wollemi pine.

Chapter 8

Captive breeding and reintroduction

Endangered species in captivity are managed to maximize retention of genetic diversity over long periods, usually by minimizing kinship. Captive populations can provide individuals for reintroductions, but their success is reduced by inbreeding depression, loss of genetic diversity and genetic adaptation to captivity

Terms

Coancestry, *ex situ* conservation, kinship, mean kinship, minimizing kinship, reintroduction, studbook, translocation

A selection of endangered species that have been captive bred or propagated and reintroduced into the wild: Mauna Kea silversword (Hawaii), California condor, black-footed ferret (North America), Przewalski's horse (Mongolia) and Arabian oryx.

Why captive breed?

Many species are incapable of surviving in their natural habitats, due predominantly to human impacts

For terrestrial vertebrates alone, it is estimated that 2000–3000 species will require captive breeding over the next 200 years to prevent their extinction. Already, 25 animal species, including Arabian oryx, black-footed ferret, California condor, Guam rail, Père David's deer, Przewalski's horse, scimitar-horned oryx, Potosí pupfish and 11 species of *Partula* snail, together with the Franklin tree and several other species of plants, have been preserved in captivity following extinction in the wild. Further, many threatened species have captive populations that act as insurance against extinction in the wild.

In this chapter we concentrate predominantly on animals, as their captive breeding is generally more complex than that for plants.

IUCN has recognized the critical contribution of captive breeding programs as they:

Guam rail

The World Conservation Union (IUCN) has endorsed captive breeding as an essential component in conservation

- establish populations in secure *ex situ* locations
- educate and engage the public on conservation issues, and provide a focus for funding-raising efforts
- provide opportunities for research on the basic biology of species, yielding knowledge that can be applied to conservation in the wild
- provide animals for reintroduction programs, where applicable

Extent of captive breeding and propagation activity

Considerable resources are being devoted to captive breeding and other *ex situ* conservation activities, but they are inadequate to conserve all needy threatened species

Approximately 1150 zoos and aquaria worldwide currently house about 1.2 million individual animals. Perhaps 5–10% of the available spaces are used for endangered species. With changes in priority, this could be expanded to breeding space for about 800 endangered species, contrasting sadly with the requirements to captive breed approximately 2000–3000 species of terrestrial vertebrates alone. As of 1989/90 245 threatened vertebrate species were being bred in captivity. Perhaps 30% of all vascular plant species are represented in the collections of the ~1500 botanic gardens worldwide.

Zoos in the twenty-first century

Zoos are now involved in a wide range of collaborative captive breeding programs

During the last 20 years, the conservation role of many zoos has expanded substantially, with active participation in a wide range of collaborative captive breeding and conservation activities. In North America, Species Survival Plans (SSPs), co-ordinated by the American Zoo and Aquarium Association, were first developed in 1981. SSPs involve co-ordinated management of all captive individuals held by co-operating institutions and are now established for many endangered species. **Studbooks** (computer databases containing the pedigree and life history of all individuals) assist recommendations on which animals should breed, with whom, how often and where. Individual institutions permit their animals to be managed under one genetic and demographic objective determined by the co-ordinator, and breeding animals are frequently interchanged to optimize

Père David's deer

genetic management. Similar programs have also been developed in many other countries and regions.

Pedigrees, individual histories, breeding experiences and health records are also collected at each zoo and maintained by the International Species Information System (ISIS) in Minnesota. ISIS provides a central repository, with (as of late 2002) 1.56 million animal records on ~10 000 species from 586 zoos and wildlife parks in 72 countries.

Potosí pupfish

Stages in captive breeding and reintroduction

Captive breeding and reintroduction may be viewed as a process involving six stages (Fig. 8.1):

Franklin tree

1. Recognizing decline of the wild population and its genetic consequences.
2. Founding a captive population.
3. Growing the captive populations to a secure size.
4. Maintaining the captive population over generations.
5. Choosing individuals for reintroduction.
6. Managing the reintroduced population (probably fragmented) in the wild.

Essential genetic issues in the first stage are the rate of decline of the wild population, the size to which it is reduced, and the resulting loss of genetic diversity and increased inbreeding prior to captive breeding. The first four issues were introduced in Chapters 4 and 5 and their management implications are discussed here.

Management during the foundation, growth and maintenance phases focuses on different priorities. During foundation, population size is small, and knowledge of the husbandry of the species is usually lacking. Management focuses on both basic research to develop husbandry techniques and on efforts to ensure reproduction of all founders.

During the growth phase, the goal is rapid reproduction and dispersion of the population to multiple facilities. The population is managed at zero population growth, through the maintenance stage at a predetermined target size. Typically, animals are not removed (e.g. for reintroduction) until the population approaches this target size.

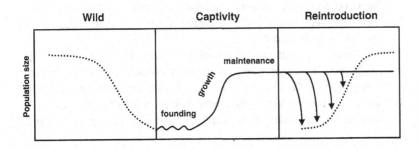

Fig. 8.1 Captive breeding and reintroduction as components of a six-stage process.

Founding captive populations

Captive populations should be founded from at least 20–30 breeding individuals. However, many such populations of endangered species have been founded with fewer individuals, with considerable loss of genetic diversity and rapid inbreeding

The founding process sets the genetic characteristics for, and ultimately affects the conservation value of, the captive population. If the population is to encompass most of the genetic diversity in the wild and minimize subsequent inbreeding, then a fully representative sample of at least 20–30 unrelated contributing founders is required. Typically, only some of the wild-caught individuals breed in captivity. For example, only 48 of 242 wild-caught golden lion tamarins entering the captive population contributed to the current gene pool. Further, two-thirds of the gene pool prior to management was derived from just one prolific breeding pair. In contrast, all 14 California condor founders successfully bred.

The estimate that 20–30 founders should be sufficient comes from equation 4.1. The proportion of heterozygosity retained is $[1 - (1/2N_e)]$. Thus, even 10 unrelated contributing founders capture 95% of the heterozygosity in an outbreeding species, while 30 founders capture over 98%. If founders come from small populations, they are likely to be related, reducing the effective numbers. Molecular techniques can be used to determine relatedness among founders. For example, the 14 California condor founders have been grouped into three related clans based on DNA fingerprint analysis.

Potential founders may come from different sources or be of unknown origin. To avoid outbreeding depression or creation of a population of undesirable hybrids, it is essential to resolve any taxonomic uncertainties, or the need for separate management units, prior to foundation (Chapter 6). Several of the founders of the 'Asiatic' lion captive breeding program were later identified as African × Asiatic lion hybrids, resulting in termination of the program after substantial resources had been expended in its development and support.

There are economic trade-offs among the number of founders, the cost of starting captive populations and the subsequent size required to maintain 90% of genetic diversity for 100 years (the current objective of captive management). With few founders, the initial cost of obtaining individuals is minimized, but larger populations are required to maintain genetic variation and avoid inbreeding. Subsequent costs will be much greater. With more founders, smaller populations sizes are needed so the initial costs are higher, but subsequent costs are reduced, yielding substantial long-term savings.

The need to acquire a solid genetic base, the low proportion of founders that reproduce, and economics all argue for founding populations with large numbers.

Unfortunately, most captive populations have been established using inadequate numbers. Some were only created when the endangered species was 'at the last gasp', such that few individuals were available. For many other captive breeding programs the few animals (or ancestors of animals) that were already in captivity were used (Table 4.1).

To avoid detrimental genetic impacts, captive populations should be established before the wild population approaches extinction. Consequently, IUCN has recommended their establishment before wild populations drop below 1000 individuals. Benefits of this strategy include (1) the ability to obtain wild founders with low inbreeding levels, (2) reduced impact of removal on wild populations, and (3) it provides time to develop suitable husbandry techniques.

> IUCN recommends that captive populations be established when wild populations drop below 1000

Growth of captive populations

The second phase of captive breeding is to rapidly multiply the population to the size predetermined by the genetic and demographic objectives of the program. This also creates a demographically secure population. Genetic management is de-emphasized during this phase as it may conflict with the primary goal of rapid population growth. Offspring are usually produced from all adults, not just those most genetically valuable. Genetic management during this phase is therefore limited to avoiding pairing of close relatives, ensuring that all animals have the opportunity to breed, and that the genetically most valuable animals are placed in viable breeding situations.

> During the growth phase of captive populations, priority is on rapid expansion as opposed to intense genetic management

How is the target population size set?
For a stable population, the effective size required to meet the goal of retaining 90% of genetic diversity for 100 years, is $N_e = 475/L$ where L is the generation length in years (Chapter 5). For example, it takes an effective population size of 475 for a species with one generation a year, but only 21 to retain 90% of the genetic variation in a species, such as mountain gorillas, with a generation length of 23 years. In practice, demographic security would dictate an effective size larger than 21 for gorillas.

> The target size is the population size required to retain 90% of genetic diversity for 100 years

The founding phase has a significant impact on the required N_e, as a small founder number (bottleneck) will require a larger population overall to compensate for genetic loss during the bottleneck. The required N_e is likely to be greater than the value above. The required number of animals depends critically on the founder effects, N_e/N ratio, the generation length, and on how quickly the population increases after the bottleneck. The N_e/N ratio depends on variance in family size, sex-ratio, mating system and fluctuations in N (Chapter 4). Computer programs are available to estimate the required N_e for complex cases.

Gorilla

Genetic management during the maintenance phase

Genetic deterioration in captivity
As the population approaches its target size, management increasingly shifts to genetic issues. Minimizing inbreeding, and consequent inbreeding depression, and retaining genetic diversity are the most immediate concerns. In the longer term, accumulation of deleterious

> Captive populations deteriorate through inbreeding depression, loss of genetic diversity, accumulation of new deleterious mutations and through genetic adaptations to captivity that are maladaptive in the wild

mutations may become a threat. Genetic adaptation to captivity (discussed later) can reduce the success rate of programs to reintroduce captive-bred organisms to the wild.

Inbreeding and inbreeding depression

Many captive populations of threatened species are too small to avoid inbreeding depression within a few generations

For many endangered species, the effective sizes of closed captive populations are so small that they will suffer inbreeding and inbreeding depression over relatively short time spans (Chapters 4 and 5). For example, the endangered Przewalski's horse has an inbreeding coefficient of about 20%, despite intensive genetic management. Individual zoos can frequently keep only a few individuals of a particular species. If just two pairs were maintained, without interchange with other institutions, then inbreeding would accumulate rapidly and the colony would be at serious risk of extinction within about five generations. To minimize such problems, the total captive population of an endangered species is often managed as a single unit, with interchange, under regional (e.g. SSPs), or global management plans (e.g. golden lion tamarin).

Despite exchange of individuals or gametes, pooled populations are still not large. The SSP programs have a mean N_e of 41, smaller than sizes where inbreeding depression due to finite size occurs. At such sizes, populations will develop an inbreeding coefficient of 26% over 25 generations, an average F greater than that arising from one generation of full-sib mating.

Loss of genetic diversity

Captive populations lose genetic diversity due to bottlenecks at foundation and due to subsequent small population sizes

Captive populations of threatened species lose genetic diversity due to the bottleneck at foundation, small subsequent population size and because $N_e < N$ (Chapter 4). Consequently, loss of genetic diversity is minimized by using an adequate number of founders, minimizing the number of generations by breeding from older animals, or using cryopreservation, and maximizing both the population size and N_e/N ratio.

Current genetic management of captive populations

Captive populations are managed to minimize loss of genetic diversity in captivity

Since founders are presumed to be representative of the genetic diversity in the wild population (frequently a dubious presumption), the goal of management is to minimize any changes to the founder gene pool from one generation to the next, in other words to 'freeze' evolution in the captive population.

The common target of management is retention of 90% of genetic diversity, implying a concomitant increase in inbreeding coefficient of 10%, over 100 years. Unfortunately, even such goals are unattainable in many populations, due to small founder numbers and/or space limitations. Consequently, the genetic objective is often compromised.

Maximizing N_e/N

Captive breeding resources are critically limited while the number of species requiring salvation by captive breeding is increasing. It is therefore important to maximize the N_e for each species, whilst using the minimum number of individuals. This can be achieved using the following protocols:

- Equalizing family sizes so that $N_e \sim 2N$ (Chapter 4). This management doubles the effective captive breeding resource and is being applied in practice.
- Equalizing the sex-ratio of breeders, i.e. avoiding harems where possible (Chapter 4). This is often difficult to achieve due to male–male aggression and other social system restrictions.
- Equalizing population sizes across generations (Chapter 4). Following the foundation and growth phases, captive populations are typically maintained at relatively stable sizes.
- Maximizing generation length. This may be done by breeding from older animals, or using cryopreservation, but these procedures are not widely practised.

The benefits of genetic management of captive populations in terms of increasing the N_e/N ratio can be very large. N_e/N ratios average approximately 0.1 in unmanaged populations, but ratios of 2 can be achieved in laboratory experiments by equalizing family sizes, sex-ratios and numbers in different generations.

In captive populations of endangered species, N_e/N ratios of about 0.3 appear common, higher than for wild populations because of control over breeding. While there is still considerable room for improvement, practical considerations (mate incompatibility, compliance with veterinary and other management recommendations and stochastic factors) will probably always limit maximization of N_e/N.

Minimizing kinship

Individual founders typically make very unequal contributions to subsequent generations, for example in the golden lion tamarins previously mentioned. Minimizing kinship is one method for reducing these inequalities. Using computer simulations, Ballou & Lacy demonstrated that, from a variety of management procedures, minimizing kinship was the best means for retaining genetic variation. In brief, this procedure chooses as parents those individuals with the lowest relationship to the overall population. When applied to unrelated founders, it equates to equalization of family sizes.

> The recommended genetic management regime is to choose parents that minimize kinship

The **kinship** (or **coancestry**) of two individuals is the inbreeding coefficient of their hypothetical offspring. Methods for computing inbreeding coefficients are given in Chapter 4. We must also compute the kinship of an individual with itself, as this enters into mean kinship below. This is equivalent to the inbreeding coefficient resulting from self-fertilization, namely $\frac{1}{2}(1 + F_{individual})$.

> The kinship of two individuals is the probability that two alleles, taken one from each at random, will be identical by descent

The **mean kinship** for individual i (mk_i) is the average of kinship values for that individual with every individual in the population, including itself

$$mk_i = \sum_{j=1,}^{n} \frac{k_{ij}}{N}$$ (8.1)

where k_{ij} = kinship between i and j, and N is the number of individuals in the population. Example 8.1 illustrates the computation of mean kinship for the named individuals in the pedigree.

Example 8.1 | Computation of mean kinships

The kinships and mean kinships for the named individuals in the pedigree below are:

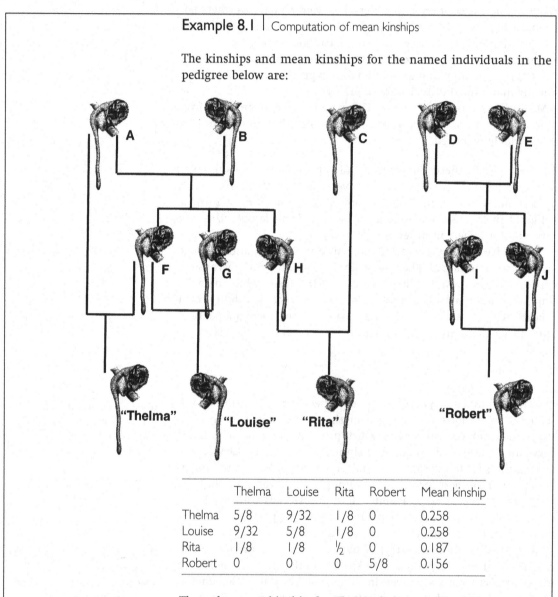

	Thelma	Louise	Rita	Robert	Mean kinship
Thelma	5/8	9/32	1/8	0	0.258
Louise	9/32	5/8	1/8	0	0.258
Rita	1/8	1/8	1/2	0	0.187
Robert	0	0	0	5/8	0.156

Thus, the mean kinship for Thelma is (5/8 + 9/32 + 1/8 + 0)/4 = 0.258.

Note that all animals, except Rita, are inbred, and their kinships with themselves are therefore greater than 0.5.

The average mean kinship for a populations ($\overline{mk}$) is the expected inbreeding coefficient of the next generation with random mating. Consequently, the relationship between average mean kinship and genetic diversity (H_t/H_0) is:

$$\overline{mk} = 1 - \frac{H_t}{H_0} \tag{8.2}$$

Consequently, if kinship is minimized, heterozygosity is maximized.

Individuals with low mean kinship are the most valuable. They have fewer relatives in the population than individuals with high mean kinship and therefore carry fewer common alleles. For the pedigree in Example 8.1, Robert has the lowest mean kinship and he is the most valuable animal in his generation as he is unrelated to Rita, Louise and Thelma. Conversely, Thelma and Louise share most genes with the rest of the population, and are least valuable genetically. Under a mean kinship breeding program, animals with lower mean kinship are given breeding priority, and we would increase the contribution of genes from Robert and decrease those of Thelma and Louise.

Applying mean kinship breeding strategies

When applying minimizing kinship in a breeding program, those individuals with lowest mean kinships are chosen to be parents. Two additional considerations are required to determine specific matings. First, matings between individuals with very different mean kinships are avoided as they limit management options in future generations. For example, if a valuable individual is mated to one of low value, increasing the contribution of the under-represented individual also increases the contribution of its over-represented mate. Second, mating of close relatives is avoided to minimize inbreeding. For example, two sibs with low mean kinships should not be mated to each other. In complex pedigrees, mean kinship and inbreeding are calculated using computer software.

Captive management of groups

Complete pedigrees are generally not available for populations maintained in groups. Species such as chimpanzees and many ungulates breed best in multi-female, multi-male groups, so paternities are unknown. Consequently, effective population sizes, inbreeding and loss of genetic diversity cannot be accurately predicted and minimizing kinship cannot be used.

For group breeders, strategies have been devised to minimize inbreeding by exchanging individuals among groups on a regular basis (Box 8.1). These procedures are similar to the equalization of family sizes, but applied to groups rather than to individuals. Contributions of groups to be parents of the following generation are equalized and

Mean kinship for a population is inversely related to genetic diversity

Where pedigrees are unavailable, management of groups by maximum avoidance of group inbreeding is recommended individuals are regularly exchanged among groups. In the first generation in Box 8.1, males are moved one group in a clockwise direction. In the second generation, males are moved two groups in a clockwise direction. In this example, inbreeding cannot be avoided after the third generation. Genetic management of groups is not widely used in captive breeding programs.

<hr>

Box 8.1 | **Low intensity genetic management for groups using maximum avoidance of inbreeding** (Princée 1995)

Maximum avoidance of inbreeding (MAI) in pedigreed populations involves equalizing family sizes and a mating system that avoids mating between relatives for as long as possible. Application of MAI to unpedigreed groups is illustrated with eight groups in the two figures below.

Maternal and paternal lineages are indicated with letters AB, CD, ..., OP. Boxes represent breeding groups. Arrows indicate transfer of males from natal groups to host groups. Bloodlines of males are shown near the arrows. The operation of the MAI system can be illustrated by following group I. In generation one, males are moved to the next group in a clockwise direction. Males of group I (bloodlines AB) are moved to group 2. Male offspring of group 8 (bloodlines OP) are moved into group I.

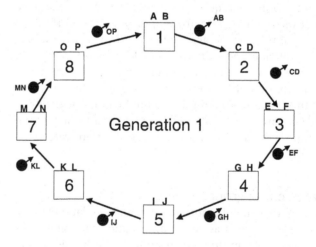

In the second generation, male offspring are moved two groups in a clockwise direction. This is necessary, as moving males by one group would result in inbreeding. For example, moving males of group I (bloodline ABOP) to group 2 (ABCD) is akin to cousin mating. Moving males by two groups results in group I males (ABOP) being mated to group 3 females (CDEF), thus avoiding inbreeding, i.e. no original breeding group is genetically represented in both males and females.

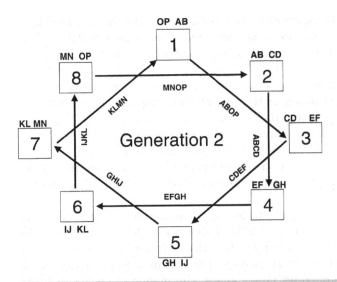

OP AB

1

MN OP

8

AB CD

2

MNOP

ABOP

KLMN

Generation 2

KL MN

7

IJKL

ABCD

CD EF

3

GHIJ

CDEF

EFGH

IJ KL

6

EF GH

4

GH IJ

5

In the third generation, males are moved four groups in a clockwise direction, I to 5, 2 to 6, 3 to 7 and 4 to 8, again avoiding inbreeding. However, inbreeding can no longer be avoided after the third generation as all groups contain bloodlines ABCDEFGHIJKLMNOP. Exchange of individuals in the fourth generation reverts to that used in the first generation and the cycle is repeated.

Ex situ conservation of plants

Genetic management of captive plant populations is not nearly as complex as in animals. Many plants can be stored as seeds, and some can be cryopreserved. Consequently, they progress through fewer generations in captivity and suffer little inbreeding, loss of genetic diversity, mutational accumulation or genetic adaptation due to captive propagation. However, preservation of seeds is not successful for about 15% of, mainly tropical, species that lack seed dormancy.

The major issue for most plants is the sampling regime involved in collection of specimens for captive propagation or storage. Two critical features are the size of samples and ensuring that collected material is representative and encompasses the full genetic diversity of the species. Representation is a greater concern for plants than for animals, especially for selfing species, where a higher proportion of genetic diversity is distributed among populations.

Generally, recommended sampling regimes involve collection of 1–20 seeds from each of 10–50 individuals from each of five separate threatened plant populations.

> Most *ex situ* conservation of plants is achieved by seed storage with occasional captive propagation to regenerate seed. Thus, the emphasis is on germplasm collection to maximize genetic diversity

Management of inherited diseases

Populations derived from a small number of founders have a high probability of expressing some genetic disorders at greatly increased frequencies. We have already mentioned undescended testes in

> Populations of endangered species initiated with few founders are likely to exhibit genetic disorders at relatively high frequencies

Florida panthers, dwarfism in California condors, hernias in golden lion tamarins, hairless offspring in red-ruffed lemurs, malabsorption of Vitamin E in Przewalski's horse and missing testicles in koalas (Chapter 2).

Initially it may appear that inherited defects should be eliminated. However, this comes at the cost of removing individuals from the breeding population, thereby reducing N_e, leading to loss of genetic diversity, increased inbreeding and slower population growth. Consequently, detailed cost–benefit analyses must be undertaken. The management options are to (a) attempt to eliminate the deleterious allele, (b) minimize phenotypic frequency of the condition, or (c) ignore it.

The impact of programs to remove deleterious alleles depends on the mode of inheritance of the defect. If it is dominant, then it can be eliminated simply by removing affected individuals. However, most inherited defects are recessive and most individuals carrying the allele are heterozygotes. Consequently, elimination of a recessive defect is very difficult, as carriers are phenotypically normal. Exclusion of suspected carriers from the breeding stock, based on pedigrees, is likely to eliminate an unacceptably large proportion of the population, along with their genetic diversity and reproductive potential. Management of a recessive genetic disorder is exemplified by chondrodystrophy in California condors (Box 8.2).

California condor

Box 8.2 | Managing a genetic disease in the California condor (after Ralls et al. 2000)

Here we discuss practical genetic management of the autosomal recessive genetic disorder chondrodystrophy.

The allele causing the condition had an estimated frequency of 0.17, assuming Hardy–Weinberg equilibrium (Chapter 2). As all affected individuals die, natural selection is expected to reduce the allele frequency from 0.17 at hatching to 0.145 after selection (Example 3.2). With random mating, the expected frequency of affected individuals in the next generation is $q_1^2 = 0.145^2 = 2.1\%$.

However, management that avoids matings between known and suspected carriers should further reduce the phenotypic frequency. In subsequent generations, it will eventually be impossible to avoid matings between all individuals that may be carriers. However, the frequency of affected hatchlings can still be kept low by splitting up pairs that have an affected offspring and pairing them with other individuals with a low risk of being a carrier.

A detailed examination of the condor pedigree in 2001 indicated that it would be possible to reduce the frequency of the gene to zero in the next generation. However, exclusion of all potential carriers (77 of 146 condors) from the current reproductive pool imposed far too high a price. The low frequency of homozygous affected individuals and the penalties in terms of demography, loss of genetic diversity and increased future inbreeding led to rejection of the option.

The adopted recommendation was to separate the pair that had produced the four affected chicks and to mate them with other individuals presumed to be totally free of the *dw* allele. In future, pairings will be made to minimize the risk of producing affected individuals, while simultaneously maximizing retention of genetic diversity through managing by mean kinship.

Reintroductions

An important role of captive populations in some conservation programs is supply of animals for reintroduction into the wild to re-establish or supplement existing populations. While such programs are in place for many species, conditions are frequently not amenable for most, as wild habitat has been destroyed, or the original threatening processes still exist. Nevertheless, many captive breeding programs aim to maintain sufficient genetic diversity and demographic viability over the long term to retain this option.

Many captive populations are managed to retain the option of eventual reintroduction to the wild

While some reintroductions have been carried out after only brief periods in captivity (e.g. black-footed ferret, California condor, Lord Howe Island woodhen), many may have to occur after long captive histories. The scenario envisaged is that humans will undergo a demographic transition within about 100 years, resulting in population decline. This may release wild habitat for the reintroduction of endangered species.

The case of the golden lion tamarin represents, arguably, the most extensively managed captive breeding and reintroduction program (Box 8.3).

| Box 8.3 | Genetic management of captive, reintroduced and wild populations of golden lion tamarins (Ballou & Lacy 1995; Grativol *et al.* 2001) |

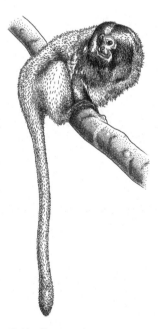

Golden lion tamarins are small, arboreal, monogamous primates from Brazil. Tamarin numbers have declined and the species has become endangered since their habitat in the Atlantic Rainforest has been fragmented and reduced to less than 2% of its original area. The Golden Lion Tamarin Conservation Program is a collaboration including the Smithsonian National Zoological Park, the Golden Lion Tamarin Conservation Association and the Brazilian Government. It is the largest global captive breeding and reintroduction program and has developed a multidisciplinary approach to preserving the species and its habitat. It integrates captive breeding, reintroduction, translocation, studies on the ecology of wild tamarins, habitat restoration and community conservation education.

The captive population consists of about 500 individuals located in 140 zoos worldwide. The concept of managing by mean kinship was originally developed for genetic management of this complex population. Although descended from 48 founders, about two-thirds of the genes in the population had derived from just one prolific breeding pair, prior to genetic management (see pedigree below).

Golden lion tamarin

However, due to subsequent careful genetic management, the level of inbreeding in the population is now only 1.9%.

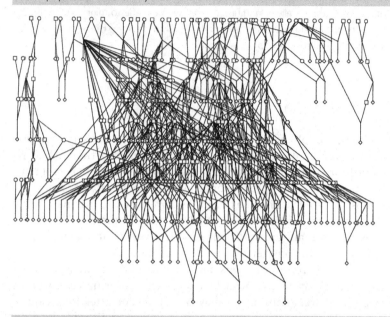

The wild population consists of about 600 individuals dispersed among 11 isolated refugia (see map below), the largest being Poço das Antas Biological Reserve containing about 350 tamarins. Translocation of animals from the most severely threatened populations to a newly established reserve established a second large protected population. Molecular analyses indicate that small populations have lost genetic diversity and become genetically differentiated from each other. This information will be used to design a program of regular translocations among fragments to minimize inbreeding and maximize the effective size of the entire population.

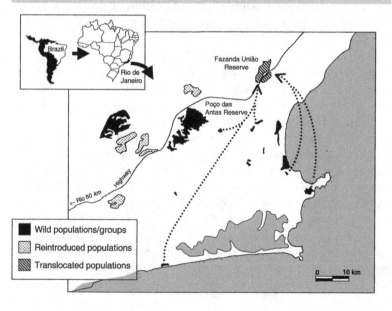

A reintroduction program was initiated in 1983 to re-establish populations in areas of their former range. Over 145 tamarins have been released, and the reintroduced population has flourished, reaching over 400 animals in 2002 with a growth rate of 25% per year in recent years.

The reintroduced population is genetically managed in a similar way to a zoo population. Weekly monitoring tracks parentage, birth and death dates, and migration events among groups. These are recorded in a studbook database and used to identify genetic relationships in the population. Individuals for reintroduction are carefully selected. From pedigrees of both the reintroduced and captive populations, release of animals that are (a) genetically valuable to the captive population, or (b) closely related to previously reintroduced animals is avoided. Similar approaches are applied to the black-footed ferret and California condor reintroduction programs.

Genetic changes in captivity that affect reintroduction success

So far, we have considered genetic deterioration in captive populations during the decline of the original natural population, and during founding, growth and maintenance of the captive colony. This deterioration results from stochastic processes in small populations and reduces the probability of success of reintroduction programs (Chapter 4). There are, however, additional and frequently overlooked, deterministic factors at work. Translocation from the wild to captivity represents a dramatic environmental change, altering the selective pressures acting upon the population. Previous natural selection may be relaxed and the population will evolve to adapt to its new environment.

The stochastic and deterministic factors have different consequences, operate over different time scales, and have different relationships with population size (Fig. 8.2). Inbreeding depression, loss of genetic diversity and mutational accumulation are all more severe in smaller than in larger populations. Conversely, genetic adaptation to captivity is more rapid in larger than smaller populations (Chapter 4). While this is beneficial in captivity, its deleterious effects are only felt when the population is returned to the wild. All detrimental changes in captivity are likely to be more severe when populations are reintroduced to harsher conditions in the wild (Fig. 8.2). Experimental results confirm the basic predictions in Fig. 8.2.

> Inbreeding depression, loss of genetic diversity, genetic adaptation to captivity, relaxation of natural selection and accumulation of new deleterious mutations reduce reintroduction success

Genetic adaptation to captivity

Genetic adaptation to captivity has been recognized since Darwin's time, but until recently, it has been considered only a minor problem in captive breeding. There is now evidence that it can be a major threat to the success of reintroductions.

When wild populations are brought into captivity, the forces of natural selection change. Populations are naturally or inadvertently selected for their ability to court, mate and breed in the captive environment. Flightiness in animals such as antelopes, gazelles,

> Genetic adaptations to captivity typically reduce reproductive fitness when populations are returned to the wild

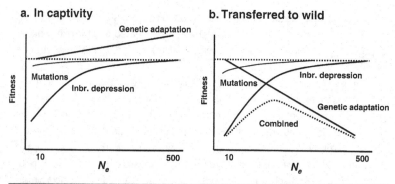

a. In captivity

b. Transferred to wild

Fig. 8.2 Genetic deterioration in captivity and its impact on reintroduction success, predicted relationships of reproductive fitness with effective population size (N_e) due to impacts of inbreeding depression, accumulation of deleterious mutations and genetic adaptation to captivity. Reproductive fitness under captive conditions is shown on the left, and for captive populations translocated to the wild on the right. The straight dotted lines are the fitness of a large wild population. The *combined* curve is the cumulative impact of inbreeding depression and genetic adaptation to captivity. A long-term time frame is being considered (~50 generations).

wallabies and kangaroos is naturally selected against when animals die after running into fences, etc. Tameness is actively favoured by some keepers. Prey capture and predator avoidance are no longer selected for while veterinary care and hygiene remove most selection for disease and parasite resistance.

Genetic adaptation to captivity has been found in all outbreeding species where it has been studied. For example, wild rats showed about a three-fold increase in reproductive output over 25 generations in captivity.

Minimizing genetic adaptation to captivity

Genetic adaptation to captivity can be reduced by:

Genetic adaptation to captivity can be minimized by reducing the number of generations in captivity, the amount of selection, and the size of captive populations

- minimizing number of generations in captivity
- minimizing selection in captivity
- minimizing genetic variation within populations
- maximizing the proportion of wild immigrants and the frequency and recency of migrant introduction
- maximizing generation length.

Minimizing time in captivity has both genetic and non-genetic benefits. Low numbers of generations in captivity have been utilized for Lord Howe Island woodhens, black-footed ferrets and California condors. Conversely, Przewalski's horse has been in captivity for most of the twentieth century and Père David's deer for hundreds of years.

Cryopreservation has the potential to be highly effective in minimizing generations in captivity. However, it is currently available only for a small number of species, generally those closely related to domestic animals. Breeding from older animals can also extend the generation interval, but has considerable management difficulties and risks.

Selection can be minimized by creating an environment as similar as possible to the natural habitat. However, duplicating a natural environment is extremely difficult and hard to justify in captive breeding, as the first priority is to establish secure and viable populations of the threatened species.

Population fragmentation as a means for minimizing genetic adaptation to captivity

When future reintroduction is envisaged, the objectives are to maintain genetic diversity, avoid inbreeding depression and avoid genetic adaptation to captivity. Neither large, nor small populations are ideal for all these purposes (see Fig. 8.2). Small populations lose genetic variation and experience inbreeding depression, but minimize genetic adaptation to captivity. Large populations retain genetic variation and inbreed more slowly, but adapt most rapidly to captivity.

A compromise may be achieved by maintaining a large overall population, but fragmenting it into partially isolated sub-populations (Fig. 8.3). The sub-populations are initially maintained separately

Fragmented populations with occasional exchanges of individuals reduce genetic adaptation to captivity and retain more overall genetic diversity than in a single population of the same total size

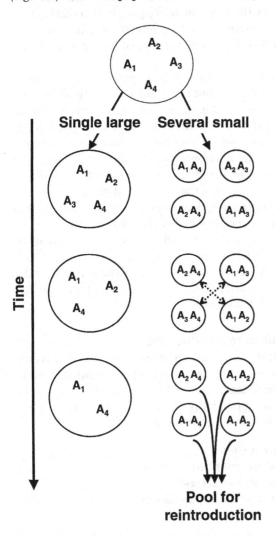

Fig. 8.3 Single large or several small: alternative genetic management options for captive populations. *Several partially isolated fragmented sub-populations, when pooled, should maintain more genetic variation, show less genetic adaptation to captivity and have a lower inbreeding coefficient than a single large population with the same total size.*

until inbreeding builds to a level where it is of concern. This will most probably be at an F of 0.1–0.2, although populations should be monitored for signs of fitness decline. Immigrants are then exchanged among sub-populations. The sub-populations are again maintained in isolation until inbreeding accumulates. This structure is expected to maintain more genetic diversity than a single population of the same total size and to exhibit less deleterious genetic adaptation to captivity. There will be low but tolerable levels of inbreeding depression.

When all sub-populations are combined (e.g. to produce animals for reintroduction) the pooled population has a lower level of inbreeding and more genetic diversity than would a single large population of the same total size. These predictions have been validated in experiments with fruit flies. A critical requirement in the use of fragmentation is that none of the sub-populations dies out.

This structure satisfies other requirements in captive breeding. Individual zoos and wildlife parks have limited capacity and endangered species are already kept 'dispersed' over several institutions to minimize risk from catastrophes. However, individuals are currently moved among institutions fairly frequently to create, effectively, a single large population. The fragmented structure with less frequent migration will reduce costs, the risk of injury, and disease transmission.

How could the fragmented structure be implemented? Initially it is envisaged that movement of animals among institutions for genetic management would cease, but that populations within institutions would be managed by mean kinship. When inbreeding levels reach 0.1–0.2, nearby zoos would exchange animals, for example, all breeding males. The new outbred populations would again be maintained in isolation until inbreeding rises to 0.1–0.2. Exchange of animals would then be among zoos that are more distant. The process would be similar to maximum avoidance of inbreeding with groups (Box 8.1). The number of generations between animal transfers depends on the N_e of each population, and will be more frequent for smaller populations than larger ones.

This strategy has only recently been advocated and experimentally validated and is not yet officially recommended. Nonetheless, it has both conceptual and practical merit.

Genetic management of reintroductions

The genetic objective in reintroduction programs is to re-establish populations with maximum genetic diversity using individuals predicted to have the highest reproductive fitness in the wild

There are many practical considerations when reintroduction is contemplated. While genetics may play only a relatively minor role in some decisions, it should not be ignored as frequently occurs. Reintroductions have a higher probability of success if all issues are considered (see below). Genetics can contribute to:

- choice of reintroduction sites
- choice of individuals, and numbers, to release
- deciding upon the number of release sites
- genetic management of released populations.

Choosing sites for reintroduction

The environment of the reintroduction site should match, as closely as possible, the environment to which the population had adapted prior to captive breeding. The site should be within the previous range of the species and, ideally be prime, rather than marginal, habitat. This will minimize the extent of adaptive evolution required in the reintroduced population. For example, the endangered northern hairy-nosed wombat is now restricted to a population at Clermont in northeastern Australia. However, genetic analysis of museum specimens revealed that a population existed at Deniliquin, some 1400 km further south, less than 100 years ago. This may therefore be a suitable site for reintroduction.

Choosing individuals to reintroduce

All other things being equal, healthy individuals with high reproductive potential, low inbreeding coefficients and extensive genetic diversity should be chosen for reintroduction. However, the impact on the secure captive population of loss of valuable animals must also be considered.

> Reintroduction programs should use healthy individuals with as much reproductive ability and genetic variation as possible

When an individual is reintroduced, its genetic diversity is added to the wild population, but removed from the captive population. Since survival in the natural habitat is expected to be much lower than in captivity and the introduced individual may not breed, it is undesirable to deplete the captive population of genetically valuable animals to benefit the wild population. Conversely, an otherwise ideal reintroduction candidate may be closely related to animals previously released and its reintroduction may actually increase the level of inbreeding in the wild population.

> When choosing animals for reintroduction, their genetic benefit to the reintroduced population must be weighed against their loss from the captive population

The interests of the two populations frequently conflict, as illustrated for golden lion tamarins in Fig. 8.4. Individuals (points) in quadrant A are those that would benefit the reintroduced population, but be harmful to the captive population. These are genetically valuable captive animals with few relatives in the wild population. Individuals in quadrant B will benefit both the wild and the captive population. These are genetically over-represented captive animals with few reintroduced relatives. Release of individuals in quadrant C would be detrimental to both populations, as they are valuable in the captive population but have many reintroduced relatives. Individuals from quadrant D are over-represented in both populations. Their release is beneficial to the captive population, but detrimental to the reintroduced population.

Because of the high risks of mortality commonly associated with release of captive-bred animals, individuals of type D should be used initially. This has been practised with golden lion tamarins, California condors and black-footed ferrets. When survival and reproduction in the reintroduced population has improved, more valuable animals may be added until the full range of genetic diversity in the captive population is represented in the wild. Survival and

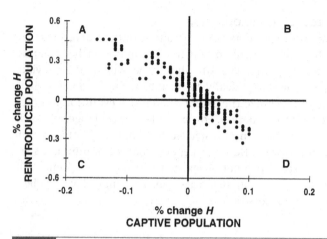

Fig. 8.4 Changes in genetic diversity in the captive and reintroduced populations resulting from releasing individual golden lion tamarins. Each point shows how reintroducing that individual will affect the heterozygosity (*H*) of the wild population and how its removal will affect the heterozygosity in the captive population.

reproduction of individuals in the reintroduced population must be carefully monitored.

Choice of individuals to reintroduce where several populations exist

Several captive populations may be able to supply individuals for reintroductions, especially in plants. The options are for individuals to be taken from:

- a single population
- a combination of populations
- crosses among populations.

The decision requires evaluation of the probability of success from each option.

A single population may be selected for species with strong local adaptation, provided a captive population was derived from the environment where the reintroduction is planned. This presumes accurate records of the source of each population, and that the environment has not changed. The latter presumption is often untrue, where the initial endangerment arose from environmental deterioration. If the environment has changed, reintroduction success will be improved by using individuals representing a combination of populations or, ideally crosses among populations.

How many reintroduced populations should be established?

Where several suitable sites and ample captive-bred individuals are available, a number of reintroduced populations should be established to maximize the numbers of individuals. This minimizes loss of genetic diversity and will minimize inbreeding if individuals are translocated among the different sites. Establishment of several

Where several captive populations exist, individuals for reintroduction may come from a single population, a pool of populations, or from crosses among different populations

There are genetic and non-genetic advantages in establishing several reintroduced populations

populations reduces the risk of extinction due to catastrophes and environmental stochasticity. For example, at least 10 reintroduced populations of the black-footed ferret have been recommended.

Genetic management of reintroduced populations

As the reintroduced population becomes self-sustaining, the more valuable animals of type B and then of type A should be released (Fig. 8.4). Eventually the wild population will gain all the genetic diversity available from the captive population, inbreeding will be minimized and reproductive fitness increased. Subsequent genetic management follows that for wild populations, discussed in Chapter 7.

> Once populations are reintroduced and self-sustaining, the full range of genetic diversity should be added to them

How successful are reintroductions?

Reintroduction of threatened species is a complex process requiring considerable understanding of species' biology and ecology. Even introductions of widespread and successful pest species, such as rabbits and European starlings, have initially failed.

> Reintroduction is a complex process with a relatively low success rate to date

Beck and colleagues considered only 11% of reintroduction programs to be successful. Their criterion was that reintroduced populations reached at least 500 individuals, free of human support. Successes included bison, Arabian oryx, Alpine ibex, bald eagle, peregrine falcon, Aleutian goose, and Galápagos tortoise. These projects were typified by release of more animals and for longer times. Successful projects also more frequently included local employment and community education programs than other projects.

The relatively low success rates for reintroductions are not surprising since some programs may have not been allowed sufficient time to succeed. Further, some programs have been somewhat cursory. Animals have been reintroduced and left to fend for themselves without training for coping with the wild. Reintroduction success should improve substantially as the science behind it advances. When evaluating the success rate for reintroductions, we should note that the success rates of recovery efforts in the wild are also low.

Bison

Case studies in captive breeding and reintroduction

Arabian oryx

The Arabian oryx was driven to near extinction in the wild because of hunting with increasingly sophisticated weapons. The species was saved by capture of the last wild animals in 1962, followed by a captive breeding program based on 10 founders. At the Phoenix Zoo, Arizona, there was a steady increase in numbers and implementation of a genetic management program. With funding and co-operation from the Sultan of Oman, a successful reintroduction program was carried out. Ten animals were initially introduced into pens, followed by gradual release in their native range. Local tribesmen were employed in management and tracking of the oryx. Released animals showed good survival and reproduction, with relatively normal behaviour. By 1995, there were approximately 280 animals in the wild, ranging over 16 000 km^2 of desert.

Unfortunately, poaching recurred and the wild population was so depleted that most individuals have been returned to captivity. Another reintroduction program has been undertaken in two fenced sites in Saudi Arabia, where poaching is currently less of a threat. These populations numbered over 600 animals in 2002.

Microsatellite analyses have revealed loss of alleles due to the founder bottleneck, errors in the pedigrees and the occurrence of simultaneous inbreeding and outbreeding depression in the species.

Black-footed ferret

The black-footed ferret was reduced in numbers through habitat loss and especially from vigorous efforts by cattle ranchers to exterminate their prey, prairie dogs. The last wild population was discovered in Wyoming in 1981 and a captive breeding program began in 1985. When distemper spread through the population, the last wild individuals were captured. The captive breeding program was established from 18 individuals (including two mothers with their litters), that contributed unequally. By 2002, the population had descended from only seven founders. The captive population has increased to about 250 adults, and has been subject to genetic management based on minimizing kinship.

A reintroduction program commenced in 1991 with surplus animals from the captive population. Since then, all females in the captive population are routinely bred to produce animals for reintroductions. Currently reintroduction has established seven populations. Between 1991 and 1999, about 1185 ferrets were introduced into the wild. In 2000, over 85 litters were born in the wild, but only two of the reintroduced populations are considered successful. The eventual objective is to establish 10 populations with a total of at least 1500 animals.

California condor

The California condor was reduced in population size due to habitat loss, DDT pollution and lead poisoning (from eating shot wildlife carcasses). After much controversy, the last birds in the wild were captured and a captive breeding and reintroduction program, based on 14 founders, was instituted at the San Diego Wild Animal Park, Los Angeles Zoo and the World of Birds. Captive numbers have steadily increased and a limited number of releases have occurred. The major initial concern has been to increase the size of the population as rapidly as possible. Reintroduction to the wild in California initially met with limited success. Deaths occurred from lead poisoning, collisions with electricity powerlines and poisoning due to ingestion of antifreeze from vehicle radiators. Reintroductions to the Grand Canyon and Baja California in Mexico have been undertaken to avoid problems associated with proximity to humans. These birds are provided with supplementary feeding. At the end of 2002, the entire species consisted of 206 individuals, 113 captive birds, 50 wild birds in California, 37 in Arizona and 6 in Baja.

SUGGESTED FURTHER READING

Frankham, R., J. D. Ballou & D. A. Briscoe. 2002. *Introduction to Conservation Genetics*. Cambridge University Press, Cambridge, UK. Chapters 17 and 18 have extended treatments of these topics, plus references.

Ballou, J. D., T. J. Foose & M. Gilpin. 1995. *Population Management for Survival and Recovery: Analytical Methods and Strategies in Small Population Conservation*. Columbia University Press, New York. Advanced treatment of topics relating to captive breeding and reintroduction; see especially the chapters by Ballou & Lacy on use of minimize kinship in the captive management, and by Princée on genetic management of groups.

Botting, D. 1999. *Biography of Gerald Durrell*. HarperCollins, New York. An entertaining popular biography of a highly influential figure who encouraged leading zoos to be involved in conservation of endangered species.

Falk, D. A., C. I. Millar & M. Olwell. 1996. *Restoring Diversity: Strategies for Reintroduction of Endangered Plants*. Island Press, Washington, D.C. Wide-ranging coverage of plant reintroductions with several case studies.

Gipps, J. H. W. 1991. *Beyond Captive Breeding*. Oxford University Press, New York. Covers a wide variety of issues related to reintroductions.

Olney, P. J. S., G. M. Mace & A. T. C. Feistner. 1994. *Creative Conservation: Interactive Management of Wild and Captive Animals*. Chapman & Hall, London. Considers many of the issues dealt with in this chapter and has several case studies.

Chapter 9

Molecular genetics in forensics and understanding species biology

Molecular genetic analyses contribute to the conservation of species by aiding detection of illegal hunting and by resolving important aspects of species biology. Coalescence and gene tree analyses provide useful tools for understanding many of these factors

Terms

Assignment test, biparental inbreeding, clade, coalescence, forensics, gene trees, haplotype, haplotype network, mismatch analysis, neutral theory, phylogeography, selective sweep

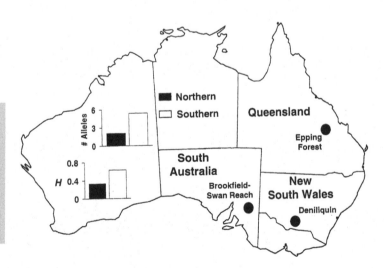

Critically endangered northern hairy-nosed wombat along with locations for its current population at Epping Forest, an extinct population at Deniliquin, a site for the southern hairy-nosed wombat, and levels of genetic diversity in the two species

Forensics: detecting illegal hunting and collecting

Poaching and illegal harvest threaten a wide variety of species, especially large cats, bears, elephants, rhinoceroses, parrots, whales and some plants. While most countries have laws to protect them, it is often difficult to obtain evidence to convict individuals illegally taking or trading in protected species. For example, a person carrying eggs suspected of belonging to threatened bird species was apprehended at an Australian airport. They avoided prosecution by simply squashing the eggs so that they could not be identified. Molecular genetic methods can now be used in such cases to identify the origin of biological material including ivory, horns, eggs, turtle shells, meat, feathers, hair and plant materials. The US Fish and Wildlife Service has established a forensics laboratory in Oregon to provide evidence in cases involving illegal imports, exports and hunting of endangered species.

> PCR-based genetic markers can be used to detect illegal hunting or collecting

One of the more fascinating cases of molecular forensics involved meat from whales on sale in Japan and Korea (Box 9.1). Analyses of mtDNA established that some of the meat was not from minke whales, for which Japan engages in 'scientific' whaling, but from protected blue, humpback, fin, and Bryde's whales. In addition, some 'whale' meat was from dolphin, porpoise, sheep and horse. Not only was illegal whaling suspected, but consumers were being misled. Similarly, PCR-based mtDNA analyses revealed that 23% of caviar being sold in New York was mislabelled. This is a concern since most of the 27 members of the sturgeon group are endangered due to over-fishing and habitat degradation. mtDNA-based methods are also being developed to detect tiger products in Asian medicines.

Sequencing of mtDNA established that 26 poached and confiscated chimpanzees in Uganda belonged to the eastern sub-species. This identified the region where poaching was taking place, and where these animals could be reinstated into the wild. The identity of a poached endangered Arabian oryx was confirmed by microsatellite analyses.

Box 9.1	Detecting sale of meat from protected whales (after Baker & Palumbi 1996; Dizon et al. 2000)

Following many years of commercial exploitation, the numbers of most whale species collapsed. The International Whaling Commission (IWC) instituted a global moratorium on commercial whaling in 1985–86. Some IWC members have continued to hunt a few whale species (primarily minke whales) for 'scientific' purposes and the meat is sold for human consumption. Suspicions arose that protected whale species were being marketed as species that could be taken legally. At the request of Earthtrust, Baker & Palumbi developed a system for monitoring the trade using mtDNA sequencing following PCR. Their protocols reliably distinguished a variety of whale species from each other and from dolphins.

Samples of whale products were subsequently purchased in retail markets in Japan and Korea. To avoid the possibility of violating laws governing transport of endangered species, Baker & Palumbi established a portable PCR laboratory in their hotel room and amplified mtDNA from the samples. The amplified DNA was taken back to be sequenced in their laboratories in New Zealand and the USA.

Results from the initial 16 purchases are shown in the figure below. They are fitted into a mtDNA phylogenetic tree together with known whale and dolphin samples. Nine samples near the top of the tree are indistinguishable from minke whales and represent meat acquired from legal 'scientific' whaling. However, sample #19b was from a protected humpback whale, while samples # 41, 3, 11 and WS4 were from protected fin whales. Consumers were being misled, as samples #16, 13 and 28 were from porpoise and dolphins.

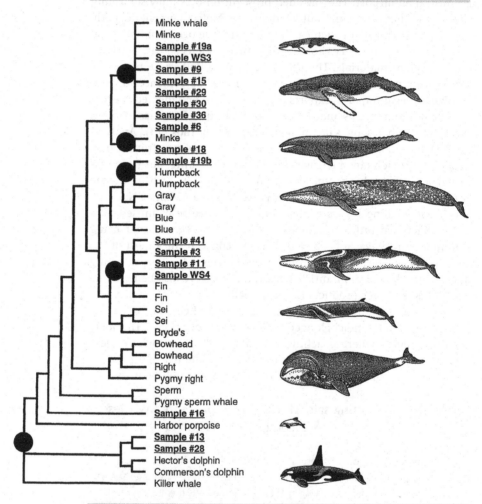

By 1999, 954 samples of 'whale meat' had been purchased in Japan and Korea and analysed by scientific groups. Of these 773 were from whales, with approximately 9% coming from protected blue, humpback, fin and Bryde's whales. Non-whale samples included dolphins, porpoises, sheep and horses. The possibility that meat from protected species had been sourced from frozen stores having been collected prior to bans on whaling cannot be excluded, but this does not apply to fresh meat. This has led to stricter controls over the distribution of 'scientifically' harvested whale meat and demands that legally harvested whales and meat stockpiled prior to whaling bans be genetically typed to monitor distribution.

Understanding a species' biology is critical to its conservation

Critical aspects of the biology of many species are unknown, as details of life histories are often difficult and time-consuming to determine directly. For example, paternities are notoriously difficult to assign without genetic data and population structures cannot be determined without analyses based on genetic markers. Introgression can typically be only suspected, but not verified from morphology. Dispersal rates may be low and unreliably estimated by direct observation.

> Molecular genetic analyses can resolve many aspects of species' biology that are critical in conservation

Molecular genetic analyses can resolve paternities, define population structures, detect introgression from other species, evaluate sources of new founders for small endangered populations, and indicate sites for reintroductions. Comparisons of DNA sequences may also be used to detect bottlenecks, migration patterns, and the demographic histories of populations. For example, life history characteristics have been determined and a potential reintroduction site identified using microsatellites in northern hairy-nosed wombats (Box 9.2). In the threatened giant tortoises on the Galápagos Islands, mtDNA and microsatellite data have been used to investigate the source of the tortoises, patterns of migration and the degree of genetic differentiation among populations, and to define conservation units. The closest living relative of the Galápagos tortoises is the relatively small Chaco tortoise on the mainland of South America. The founding events on the Galápagos Islands were generally from geologically older to younger islands, with no back migration detected, but with some islands being colonized more than once. Individual island populations were all genetically distinct, with two islands having distinct populations within them. The studies have also clarified the boundaries of conservation units.

The remainder of this chapter examines the methods used, and presents examples of practical issues that have been resolved.

| Box 9.2 | Censusing the critically endangered northern hairy-nosed wombat and inferring aspects of its biology using hair samples (Taylor *et al.* 1994, 1997; Behegeray *et al.* 2000; Sloane *et al.* 2000) |

Northern hairy-nosed wombats are restricted to a tiny location in Queensland, Australia (chapter frontispiece). They are nocturnal, fossorial, and difficult to study directly. Trapping has been used to estimate population size and sex-ratio, but is traumatic to the animals, risks possible injury or death, necessitates construction around burrows and is time-consuming and unreliable.

Using DNA from hair collected on adhesive tape on frames at the entrance to the wombat burrows, it was possible to genetically characterise the population for up to 20 microsatellite loci. This allowed each individual in the population to be identified and its sex determined by amplification of X- and Y-linked loci. The population can now be routinely censused without disturbance.

Tissue from museum skins was used to characterize an extinct wombat population at Deniliquin. Microsatellite analysis established that this population was of the northern species, rather than the southern species currently found nearer to this location. Consequently, Deniliquin should be suitable for establishing a reintroduced population as an insurance against catastrophes in the Queensland population.

The data have also been used to infer aspects of the biology of the northern hairy-nosed wombats. Individuals of the same sex sharing burrows are often relatives, but males and females sharing burrows are not close relatives. Dispersal patterns have also been deduced. The N_e/N ratio was estimated to be 0.1, based on loss of genetic diversity in comparison with the southern species. Parentage analyses based on eight to nine microsatellite loci were not particularly successful due to low genetic variation, but this will be improved with use of additional, subsequently developed loci.

Currently available methodologies and their applicability to the issues considered here are listed in Table 9.1. In general, DNA based methods are suitable for most purposes. Further, RAPDs, mtDNA and microsatellites can be analysed from non-intrusive samples and PCR (Chapter 2).

Gene trees and coalescence

Gene trees and coalescence analyses provide information on many aspects of species biology necessary for effective conservation

DNA sequences retain information on their prior evolutionary history of population sizes, population fragmentation, selection history, etc. Analyses of DNA sequence differences among individuals and populations allow us to explore evolutionary processes and demographic events in a species' past.

Coalescence and gene trees are two major methods for extracting this information. Based on sampling theory for neutral alleles (**neutral theory**), they provide a null hypothesis against which to test data and to discriminate possible reasons for deviations (Chapters 2 and 4). Moreover, coalescent methods work backwards in time and allow time dimensions (generations) to be added to the analyses. Consequently, they are more powerful than conventional analyses that use only current distributions and patterns of DNA sequence differences.

Coalescence is based on the concept that, if we trace current allelic sequences in a population back long enough through time, they **coalesce** to a single individual sequence (Fig. 9.1). Other alleles, once present in the past, have been lost by genetic drift or selection, and new alleles have been generated through mutation. The evolutionary pattern of the extant distribution of alleles at a locus can be represented as the branches of a tree coalescing back to a single ancestral allelic sequence.

Table 9.1 Applicability of available methods for genetically characterizing individuals and populations. Techniques with + can be used for the purpose specified, with several + indicating that the technique has higher utility, ? are cases where the technique is useful in only some cases, while − indicates that the technique is not useful in this context

Issue	Morphology	Chromosomes	Allozymes	mtDNA	RAPD	DNA fingerprint	Microsatellites
Non-intrusive sampling	−	−	−	+++	++	−	+++
Forensics	−	−	+	+++	++	+++	+++
Population size	+	−	−	+++[a]	+	?	++
Estimating N_e	−	−	++	++[a]	−	?	+++
Demographic history	−	−	−	++[a]	−	?	+
Detecting and dating bottlenecks	−	−	++	++[a]	++	?	+++
Detecting selection	+	+	+	+++	+	++	+++
Migration and gene flow	?	−	++	+[a]	++	++	+++
Individual identification and tracking	+	−	−	++	+	−	+++
Population structure	?	−	++	+?	++	++	+++
Phylogeography	−	−	−	+++	−	−	+++
Source populations to recover endangered species	+?	−	++	+	++	+++	+++
Introgression	+	+	++	+[a]	++	++	+++
Secondary contact	−	−	−	+++	−	+	+++
Taxonomic status	+	+++	++	++	+++	+++	+++
Sites for reintroduction	−	−	−	+	+	−	+++
Populations for reintroduction	+?	−	++	+	++	+++	+++
Reproductive systems	−	−	++	−	+	?	+++
Paternity	−	−	+	−	+	+++	+++
Founder relationships	−	−	?	−	+++	+++	++
Sources of new founders for endangered populations	+?	−	++	+	++	+++	+++
Sexing animals	?	+++	−	−	−	?	+
Detecting disease	−	−	−	++?	++	++	+
Diet	−	−	−	+++	++	++	++

[a] Can detect only female contributions.

Coalescent patterns are usually depicted using **gene trees**, which show the genealogy of the alleles in the current population. The nodes (coalescence events) and branch lengths in the tree reflect the origins and time frames involved in deriving the observed patterns. Gene trees trace the evolutionary history of the alleles (e.g. different mtDNA sequences) in the same manner as tracing the origin, or loss, of alleles

Coalescence is the analysis of the distribution and differences among DNA sequences for alleles and the events and time frames involved in developing these sequences

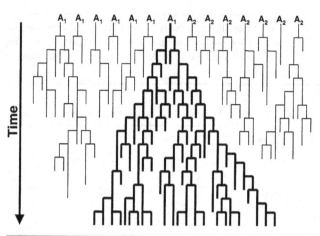

Fig. 9.1 Gene trees and coalescence: a possible history of descent of DNA sequences in a population that began at time 0 (top of figure) with 14 copies, representing two alleles. Some sequences leave one or more copies in the subsequent generation while others become extinct. Sequences present at the bottom are all descended from (coalesce to) a single ancestral copy of the A_1 allele (this lineage is shown in the heavier black lines in the figure). If the failure of the ancestral sequences to leave descendants was random, the sequences at the bottom could equally well have come from any other ancestral copy at generation 0 (after Futuyma 1998).

through pedigrees. For example, the sex-linked haemophilia allele in the royal families of Europe can be traced back to Queen Victoria of Great Britain.

Coalescence provides theory to model the survival and spread of alleles over time in the lineages of a population

The basis of the coalescence method is that DNA sequence differences among alleles at a locus retain information about the evolutionary history of those sequences. For example, two alcohol dehydrogenase alleles in *Drosophila* that differ by two bases are more closely related and diverged more recently than two alleles that differ by 11 base pairs.

Neutral theory allows us to predict the time in generations back to coalescence, thus adding a time dimension to analyses. Under neutral theory, two alleles may descend from the same ancestral allele in the previous generation with a probability $1/N_{ef}$ for mtDNA (where N_{ef} is the effective number of females), or $1/(2N_e)$ for a nuclear locus in a diploid species. Alternatively, two alleles may derive from two different alleles in the previous generation (or derive from the same allele many generations ago) with probabilities $1 - 1/N_{ef}$, or $1 - 1/(2N_e)$. This is the same reasoning used to determine loss of genetic diversity (Chapter 4). Under this neutral model of genetic drift in a diploid population with k alleles, the average time T_k back to the previous coalescent event (i.e. where there were $k - 1$ alleles) is:

$$T_k = \frac{4N_e}{k(k-1)} \text{ generations} \tag{9.1}$$

Thus, the times during which there are 5, 4, 3 and 2 lineages are $4N_e/20$, $4N_e/12$, $4N_e/6$ and $4N_e/2$ generations, respectively. The time for all alleles in the population to coalesce is $4N_e [1 - (1/k)]$

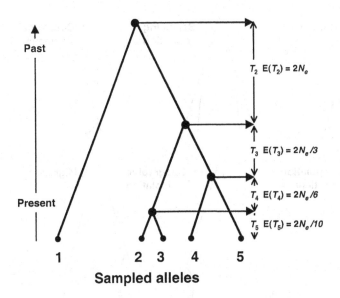

Fig. 9.2 Gene trees and coalescence times: an example of a gene tree for five alleles (after Hedrick 2000). The large circles indicate coalescent events. T_i is the length of time it takes for the i alleles present to coalesce into $i - 1$ alleles, and intervals are shown in proportion to their expected times.

generations (Fig. 9.2). Thus, the coalescence is quicker, and gene trees shorter, in smaller than larger populations.

We can immediately see an application of gene tree analysis. It can provide details about differences in historical population size for different populations, or species. Example 9.1 illustrates the calculation of coalescence times in diploid populations with effective size 50 and 100. Note that the coalescence times increase in direct proportion to population size.

Example 9.1 | Estimating coalescence times

In a population of $N_e = 50$ with three alleles, the expected time to its previous coalescence (when the population only had two alleles) is:

$$T_3 = \frac{4N_e}{k(k - 1)} = \frac{(4 \times 50)}{(3 \times 2)} = 33 \text{ generations.}$$

Thus, three alleles coalesce to two alleles on average in 33 generations in a population of size $N_e = 50$.

For $N_e = 100$ coalescence takes:

$$T_3 = \frac{4N_e}{k(k - 1)} = \frac{(4 \times 100)}{(3 \times 2)} = 67 \text{ generations.}$$

Thus, the coalescence takes twice a long in a population with twice the size.

The structure of gene trees and patterns of coalescence are strongly influenced by deviations from selective neutrality and random mating (Fig. 9.3). Different forms of selection affect the coalescence time in characteristic ways; directional selection reduces the coalescence time, while balancing selection increases it, compared to the expectation from neutral genetic drift alone. Coalescence of alleles in the

Alterations in coalescence patterns allow detection of selection, isolation among populations, and changes in population size

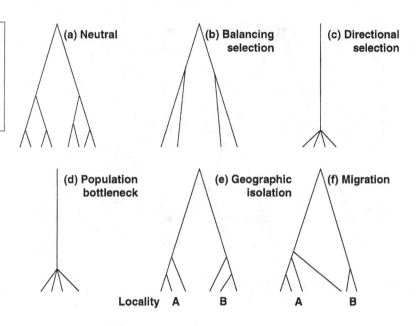

Fig. 9.3 Gene trees showing coalescence patterns for (a) neutrality, (b) balancing selection, (c) directional selection (selective sweep), (d) population bottlenecks, (e) geographic isolation, and (f) migration.

MHC or SI loci (that are subject to balancing selection) often extend back prior to speciation events.

After long periods of isolation and lack of gene flow, there are deep divisions among populations (Fig. 9.3e). Migration yields characteristic signatures when gene trees are mapped onto geographic location (Fig. 9.3f); alleles characteristic of one geographic region are found in another region. Fluctuations in population size, or population bottlenecks foreshorten coalescence time (Fig. 9.3d). Mutations generate sequence differences, slowing coalescence times.

When patterns are similar, such as those for directional selection to fixation (**selective sweeps**) and population bottlenecks, additional information is required to resolve the cause. For example, information on multiple unlinked loci allow discrimination of selective sweeps and bottlenecks (Fig. 9.3c and d); population bottlenecks affect all loci in a similar manner, while a selective sweep will affect one locus in a manner different to the behaviour of other loci.

Differences in DNA sequences, gene tree structure and coalescence rates allow us to infer details about population structure and evolution that are not easily, or less accurately, found using other techniques. Analysis of gene trees, using coalescence analysis, have been used to:

• estimate effective population sizes (using selectively neutral sequences)
• measure neutral mutation rates
• infer selection and determine its form
• determine migration events and measure migration rates
• determine phylogenetic relationships among geographically separated populations (and compare these among species to determine whether patterns are concordant)
• detect secondary contact of diverged populations

- estimate divergence times among populations
- infer changes in population size (bottlenecks, exponential growth vs. constant size)
- detect recombination in disease organisms
- reconstruct the origins and history of disease epidemics.

To this point, most coalescent analyses have used mtDNA data, as recombination is essentially absent, inheritance is maternal (in most species), and mtDNA has higher mutation rates than nuclear loci and can therefore detect effects over shorter time spans. Nuclear DNA sequences have just begun to be used. Analyses of gene trees, coalescence and phylogeographic patterns have become a discipline in their own right.

The following include applications involving gene trees and coalescence analyses in addition to other molecular approaches. As these are rapidly developing fields, with new methods appearing regularly, we are only able to describe a sample of the methods and applications.

Population size and demographic history

Population size

It is difficult to directly estimate population sizes in nocturnal, fossorial, rare and shy animal species. Minimum estimates of their population size can be made by identifying individuals using multilocus DNA fingerprints or microsatellites. DNA can be 'remotely' sampled by collecting hair, skin or faeces, and microsatellites typed following PCR amplification. For example, a minimum population size of five individuals was estimated for endangered Pyrenean brown bears as five different multilocus microsatellite genotypes were found. Of these individuals, one was a female and four were males, as distinguished using sex-specific DNA markers (see later).

Minimum estimates of population size can be obtained from the number of unique genotypes

PCR-based genetic markers can be used to identify species when using faeces to estimate population size

Scat (faeces) counts have been used to estimate population sizes for many species, e.g. bears and coyotes. However, this cannot be applied where more than one species with similar scats live in an area. DNA analyses have been used to authenticate bear scats in Europe. Similarly, five microsatellite markers and one sex-specific marker were used on seal faecal samples to identify species, individuals and sex ratios at a mixed-species 'haul-out'.

Demographic history

The distribution of the number of sequence differences between pairs of alleles (a **'mismatch'** analysis) has a characteristic shape for populations with different demographic histories (Fig. 9.4). Stable populations yield geometric distributions (Fig. 9.4a), while exponential growth is expected to generate a smooth unimodal distribution

The distribution of sequence differences between pairs of alleles can be used to distinguish between stable, growing and bottlenecked populations

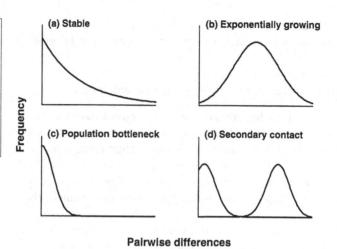

Fig. 9.4 Distributions of pairwise sequence differences between alleles in populations with different histories (after Avise 2000). (a) Population with stable size, (b) population showing exponential growth, (c) population subject to a recent bottleneck and (d) secondary contact and fusion.

(Fig. 9.4b). Bottlenecks yield either a distribution close to zero, or a bimodal distribution, depending on whether the bottleneck reduced genetic diversity, or completely removed it (so that the diversity represents mutations since that point) (Fig. 9.4c). Secondary contact of populations following long isolation yields a bimodal distribution (Fig. 9.4d). Humans exhibit a unimodal distribution characteristic of exponential growth, which accords with known human history.

Characterizing and dating bottlenecks

Signals of past population bottlenecks can be detected using molecular genetic analyses

Undocumented past bottlenecks can be detected and their severity inferred from the loss of genetic diversity. Even when there are no samples of the pre-bottleneck population, they can often be identified using information from multiple microsatellite loci.

Koalas exhibit signals of a population bottleneck, together with isolation-by-distance, in their mtDNA sequences (Fig. 9.5). Most populations from Victoria and South Australia are indistinguishable, in contrast to the diversity among populations further north. The southern populations were established by individuals translocated from island populations that had experienced bottlenecks (Box 7.3). The patterns of similarities among other mainland populations suggest isolation-by-distance, with populations from nearby localities generally being more similar than those from distant localities.

Koalas

Bottleneck effects have been measured by comparing microsatellite genetic diversity from the current populations with that from museum specimens for Mauritius kestrels (Box 4.1).

The size and duration of bottlenecks can be inferred from loss of genetic diversity

Not only can bottlenecks in population size be detected, but also the size of the bottleneck can often be inferred from loss of genetic diversity. For example, the size of the population bottleneck experienced by the northern elephant seal has been inferred from mathematical modelling of the loss of mtDNA genetic diversity. It can be accounted for by a single generation bottleneck of 10–20 effective females. However, a bottleneck of this size is not sufficiently severe to account for the complete absence of allozyme variation.

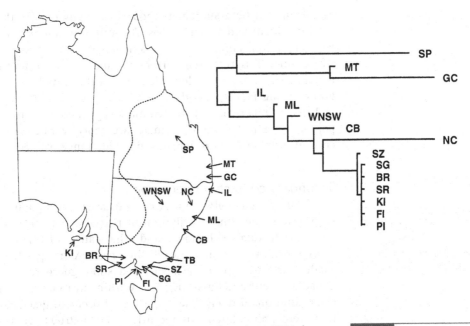

Fig. 9.5 Gene tree for koala populations, based on mtDNA sequence divergence (after Houlden et al. 1998). *Populations from Victoria and South Australia (bottom), derived mainly from bottlenecked island populations, are essentially indistinguishable. The remaining populations generally show their closest affinities with geographically adjacent populations, as expected with isolation-by-distance.*

A particularly elegant example of the use of ancient DNA is provided by a study of the nene in Hawaii. There is only a single mtDNA **haplotype** (haploid multilocus genotype) present in extant birds and, largely the same haplotype is found in museum specimens and sub-fossil bones aged 100–500 years. However, there are seven haplotypes in sub-fossil bones aged 850–2540 years. Thus, the major decline in genetic diversity did not result from Western colonization but from earlier Polynesian settlement. Modelling indicates that the population probably declined to fewer than 10 females over 50–100 generations.

The 13 species of Darwin's finches on the Galápagos Islands were thought to have diverged from the progeny of a single pair. However, MHC sequence diversity indicates that they probably diverged from a founding group of at least 30 individuals.

It has been hypothesized that the cheetah lost substantial genetic diversity due to a population bottleneck. The presumed bottleneck was estimated to have occurred about 10 000 years ago, by comparing levels of genetic variation for allozymes (low mutation rate), with mtDNA, DNA fingerprints and microsatellites (higher mutation rates).

> The timing of bottlenecks can be inferred from genetic data

Estimating evolutionary effective population size

Coalescence theory allows effective population size for females to be estimated from mtDNA as:

> Long-term effective population size can be determined using coalescence theory if mutation rates are known

$$N_{ef} = \frac{PS}{2u} \tag{9.2}$$

where PS is the average proportion of nucleotide sites that differ between random pairs of haplotypes and u is the mutation rate. The effective population size for female red-winged blackbirds in the USA was estimated by a similar method. RFLP analyses of mtDNA from 127 birds identified 34 haplotypes. The mutation rate per base is 10^{-8} per generation. From these, an estimate of 36 700 females was derived, much less than the current number of 20 million breeding females. The most probable explanation is that numbers were much lower during Pleistocene glaciation (as recently as 10 000 years ago) and that numbers have increased markedly since then from a refuge population. Effective sizes in females are typically much less than the current number of adult females in a wide range of species.

Secondary contact between populations

Secondary contact between previously differentiated populations can be inferred when the distribution of pairwise sequence differences among individuals is bimodal (and other causes of this pattern can be excluded). For example, two distinct mtDNA groups of haplotypes (**clades**) are present in each surveyed snow goose rookery in the Canadian Arctic and Russia (Fig. 9.6). Within any rookery, both mtDNA clades interbreed freely. The rookery sites occur in glaciated regions that were uninhabitable as recently as 5000–10 000 years ago. Current populations are huge, so this is not a recent drift effect. Thus, the rookeries have been established by migrants, presumably from two separate, diverged refuge populations. Heterozygote advantage can be excluded as an explanation, as it cannot operate on haploid mtDNA.

Secondary contact between populations yields a bimodal distribution of pairwise sequence differences among individuals

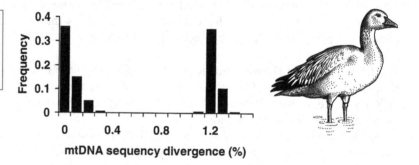

Fig. 9.6 Bimodal distribution of pairwise differences in mtDNA RFLPs for snow geese indicating secondary contact (after Avise 2000).

Gene flow and population structure

Population structure

Genetic management recommendations vary significantly depending on population structure. Populations in different habitat fragments may be totally isolated, partially isolated, effectively a single population, or a metapopulation, depending on the extent of gene flow and population extinction rates (Chapter 4). Small and totally isolated populations may experience severe inbreeding. The delineation of population structure is usually only possible using genetic data.

Analyses with genetic markers are used to determine population structure

The degree of population differentiation can be determined using F_{ST} and related measures for any type of polymorphic genetic marker (Chapter 4). More powerful and informative analyses are possible using gene trees. Population structure can be identified by mapping the sequences of different alleles onto geographic locations. The cause of genetic differentiation, restricted gene flow, past fragmentation or range expansion can then be determined. East African populations of buffalo and impala show similar F_{ST} values of 0.08 and 0.10. However, the distribution of mtDNA haplotypes over geographic locations is entirely different in the two species, as shown in the **haplotype networks** in Fig. 9.7. The distribution of Chobe haplotypes (Chobe is the most isolated location) is random in buffalo, but tightly clustered in impala. Consequently, buffalo exhibit recurrent maternal gene flow between Chobe and more northerly populations. In contrast, impala have restricted female gene flow that either reflects isolation-by-distance or isolation of the Chobe population from the northern populations.

Differentiated maternally inherited components of the genome can be revealed with mtDNA or chloroplast DNA (plants). Mitochondrial DNA analysis revealed that the vulnerable ghost bat populations in Australia exhibit marked differentiation among colonies, and no

(a) Buffalo

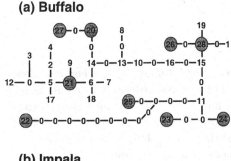

(b) Impala

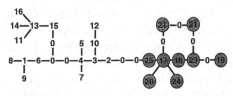

Haplotype found
only in Chobe

Fig. 9.7 Haplotype networks for mtDNA for buffalo and impala (after Templeton 1998). Each line in the network represents a single mutational change. 0 indicates a node in the network that was absent in the sample. These nodes are inferred intermediates between the two nearest neighbour haplotypes in the network that differed by two or more mutations. Haplotype numbers are those given in the original reference. *Chobe haplotypes, from the most isolated and southerly location, are tightly clustered for impala but interspersed throughout the buffalo network.*

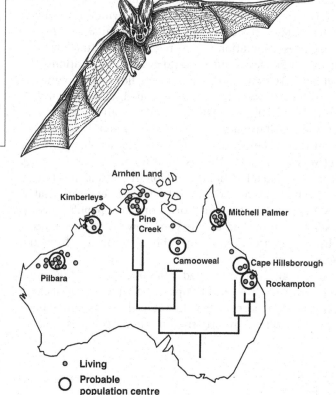

Fig. 9.8 Population structure of ghost bat populations revealed by mtDNA (after Moritz *et al.* 1996). This species lives in caves and abandoned mines and was once widespread, but has contracted northward. *There is clear differentiation among populations. No mtDNA haplotypes were shared among populations.*

mtDNA haplotypes are shared among colonies (Fig. 9.8). Further, microsatellite analyses also revealed substantial differentiation among populations indicating little if any gene flow. Consequently, extinct colonies are unlikely to be recolonized by natural migration. Moritz and colleagues recommended that each colony be managed as a separate management unit for conservation purposes.

In another application, studies in the long-finned pilot whale revealed a pod structure consisting of single extended female family lines, often containing more than 100 individuals. Neither sex of offspring disperses. However, there was not significant inbreeding, as determined using nuclear genetic markers. This apparent contradiction was resolved when members of different pods were observed to mate when they encountered each other, thereby minimizing differentiation for nuclear loci.

Dispersal and gene flow

Dispersal rates can be inferred from genetic differentiation among populations

Dispersal rates are difficult to study by direct observation, as rates may be low and long-distance dispersal too rare to measure with precision. An increasing number of studies use genetic markers to infer dispersal patterns. With no dispersal, mating will not be random, and there will be a deficiency of heterozygotes when data from all samples are pooled (Chapter 4). For example, mice maintain territories within barns and do not mate at random, resulting in overall deficiencies of heterozygotes compared to Hardy–Weinberg expectations.

Genetic studies often reveal a picture of dispersal differing from that suggested by direct observations. For example, observations of territorial North American pikas (small mammals) indicated that they rarely dispersed, and long-distance dispersal was not observed at all. Consequently, they were thought to mate regularly with close relatives. However, DNA fingerprinting studies revealed that close inbreeding was not common and that dispersal occurred over short, medium and long distances.

Dispersal patterns of female turtles have been followed using mtDNA markers. Females frequently return to lay their eggs in the beach where they hatched. Genotypes in loggerhead turtles are distinct in populations on nesting beaches in Japan (B and C) and Australia (A) and their dispersal can therefore be followed (Fig. 9.9). Populations in the North Pacific, and in the feeding grounds near Baja California, were predominantly Japanese stocks, with a minority of Australian stock. Consequently, females show fidelity to nesting sites, but not to feeding grounds. Similar studies have traced movement of endangered humpback whales.

Y-chromosome specific DNA markers allow male-specific dispersal and differentiation to be assessed in a parallel manner. In humans, the use of mtDNA and Y-specific DNA markers to estimate female and male dispersal, respectively, revealed that females disperse more

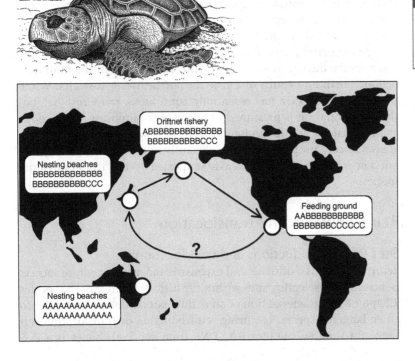

Fig. 9.9 Dispersal in loggerhead turtles based on mtDNA genotypes (after Bowen et al. 1995). A, B and C are different mtDNA types.

Driftnet fishery
ABBBBBBBBBBBBBB
BBBBBBBBBBCCC

Nesting beaches
BBBBBBBBBBBBB
BBBBBBBBBBBCCC

Feeding ground
AABBBBBBBBBBB
BBBBBBBCCCCCC

?

Nesting beaches
AAAAAAAAAAAAA
AAAAAAAAAAAAA

effectively than males. This contrasts with the prevailing view that males spread their genes more effectively due to movements associated with warfare.

Detecting immigrants

Individual immigrants can be identified from multilocus genotypes using assignment tests

Dispersal of individuals can be identified from multilocus microsatellite genotypes (**assignment tests**). Based on its genotype, an individual is assigned to the population with which it has the greatest similarity. If it is assigned to a population other than the one where it was collected, then it is presumed to be an immigrant. For example, if all individuals in geographic areas A and B have genotypes $A_1A_1B_1B_1C_1C_1D_1D_1$ and $A_2A_2B_2B_2C_2C_2D_2D_2$, respectively, then an individual in region B with the former genotype must be a migrant. An identical principle applies when populations differ in frequencies at several loci, but the computations are more complex. Assignment tests can also be used to detect hybridization and taxonomic relationships, as has been done with the red and Algonquin wolves.

Phylogeographic patterns across species

Past geological or climatic events may affect many diverse species in a similar manner

In a surprising number of cases, patterns of DNA sequence divergence across geographic regions are concordant for many species (**phylogeography**). For example, black sea bass, seaside sparrows, horseshoe crabs, American oysters and tiger beetles along the Atlantic coast in the USA show distinctively different mtDNA haplotypes than those in the same species from the Gulf of Mexico, areas separated by the Florida peninsula. These distributions were not previously recognized, nor predicted from current landforms. Such patterns apparently reflect separation of populations by major geological events, past climatic events or habitat changes resulting from climatic change. Many other cases of concordant phylogeographic patterns in distinct taxa have been found in the USA, including freshwater fish in east- vs. west-flowing rivers in the southeast. Similarly, four species of birds and a reptile display major genetic differences between the rainforest regions north and south of Cairns in northeastern Australia. While this region currently has essentially continuous rainforest habitat, the differentiation is presumed to reflect past rainforest contractions and expansions that led to long periods of isolation between the two areas. Concordant patterns across distantly related species strengthen inferences that may be only weakly supported by data from individual species.

Reintroduction and translocation

Sites for reintroductions and translocations

Potential reintroduction sites can be identified by PCR analyses on museum specimens collected from populations that are now extinct

Reintroduction is a difficult and expensive undertaking whose success is increased by selecting sites within the historical range of the species (Chapter 8). Characterization of an extinct population as belonging to an endangered species requiring reintroduction or translocation can suggest a suitable site (Box 9.2). DNA from sub-fossil bones revealed that Laysan ducks recently existed on a Hawaiian island where they are now extinct. This island may therefore be a suitable site for re-establishment.

Populations for reintroductions

A reintroduction program for the endangered shrub *Zieria prostrata* from a restricted area on the east coast of Australia was abandoned following genetic analyses. An apparently unique plant, thought to be have originated from an extinct population at a location distant from the remainder of the species, was destined to be the subject of a re-establishment program. However, it was found to be closely related to individuals from one of the extant populations, so the reintroduction plan was cancelled.

> Evaluation of candidate populations for reintroduction can be made following genetic analyses

Breeding systems, parentage, founder relationships and sexing

Species with different breeding systems (asexual vs. sexual, inbreeding vs. outbreeding, etc.) require different management, and it is vital to distinguish them (Chapter 7). Knowledge of parentage is critical in detecting inbreeding, and to verify the accuracy of pedigrees used in genetic management. Correct assignment of sex is essential so that two individuals of the same sex are not paired and so that distorted sex ratios can be recognized. Founder relationships are important in managing captive populations so that loss of genetic diversity and inbreeding can be minimized. Genetic marker analyses can provide much of this critical information.

> Genetic analyses can provide critical information on breeding systems, parentage, sex and founder relationships

Breeding systems

Plants have a diversity of mating systems, from outbreeding to self-fertilization and clonal reproduction. Some small plant populations may switch from outcrossing to selfing. Further, some species of fish, lizards and insects are parthenogenetic or self-fertilizing. As species with different mating systems typically require different genetic management strategies, it is crucial that the mating system for each threatened species be defined.

> Methods of reproduction and mating patterns can be resolved by typing mothers and offspring for multiple genetic loci

Breeding systems can be determined by genotyping mothers and offspring (Table 9.2). If all offspring contain the same genotype as the mother then reproduction is asexual (including ameiotic parthenogenesis). Conversely, if offspring contain only alleles present in the mother, but have a diversity of genotypes then they are the result of self-fertilization. Offspring containing alleles not found in the mother are the result of outcrossing.

All individuals in one population of the endangered Santa Cruz bush mallow plant on Santa Cruz Island, California were identical (indicating clonal reproduction) and different from individuals in a second population (Fig. 9.10). The endangered shrub *Haloragodendron*

Table 9.2 Determining of breeding systems using genetic markers

Breeding system	Parent genotypes		Offspring genotypes
Asexual	AB	⇒	AB
Selfing	AB × AB	⇒	AA, AB, BB
Outbreeding	AB × CD	⇒	AC, AD, BC, BD
Mixed selfing and	AB × AB	⇒	AA, AB, BB
outcrossing	AB × CD	⇒	AC, AD, BC, BD
	(heterozygote deficiency compared to outcrossing)		

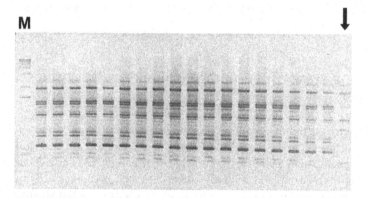

M ↓

Fig. 9.10 Clonal reproduction in the endangered Santa Cruz bush mallow (from Fritsch & Rieseberg 1996). RAPD analyses on 18 different plants from the NS(II) population (lanes 2–19) and one individual from the NS(I) population (arrow). Lane 1 is a DNA size marker (M). *All 18 bushes from the NS(II) population are identical (clones) and different from the NS(I) plant.*

lucasii exists in a very restricted range in Sydney, Australia and comprises only seven clones among 53 plants, based on allozyme and RAPD genotypes.

The extent of selfing in plants with mixed mating systems (some selfing and some outcrossing) can be estimated directly by typing maternal plants and their offspring. Selfing of homozygous maternal plants results only in homozygous progeny, while outcrossing yields heterozygotes (H) at a rate dependent upon the combined frequency of alleles not found in the homozygote (q) (Table 9.3). Thus, the frequency of selfing (S) is

$$S = 1 - \frac{H}{q} \tag{9.3}$$

In plants with mixed selfing and outcrossing, selfing rates can be determined directly by typing maternal parents and progeny using genetic markers

Heterozygous maternal plants can also be used to obtain estimates of selfing rates.

The level of inbreeding can be determined indirectly from deviations from Hardy–Weinberg equilibrium (Chapter 4). For example, the ratio of observed to expected heterozygosity, $H_o/H_e = 0.68$ in endangered round-leaf honeysuckle plants from Western Australia. Using

Selfing rates can be determined indirectly from the reduction in heterozygosity compared to Hardy–Weinberg expectations

Table 9.3 | The proportion of progeny genotypes expected from a homozygous maternal genotype (A_1A_1) and a heterozygous maternal genotype (A_1A_2) as a result of self-fertilization (S) and outcrossing (T). p and q are the frequencies of alleles A_1 and A_2 in the population

		Progeny genotypes		
Maternal genotype	Frequency of matings	A_1A_1	A_1A_2	A_2A_2
A_1A_1	S	S		
	T	Tp	Tq	
A_1A_2	S	$\frac{1}{4}S$	$\frac{1}{2}S$	$\frac{1}{4}S$
	T	$\frac{1}{2}Tp$	$\frac{1}{2}T$	$\frac{1}{2}Tq$

Source: Hedrick (2000).

equation 4.3, the inbreeding coefficient is

$$F = 1 - \frac{H_o}{H_e} = 1 - 0.68 = 0.32.$$

The selfing rate can be determined from the inbreeding coefficient, as follows:

$$S = \frac{2F}{(1+F)} \tag{9.4}$$

For round-leaf honeysuckle, the selfing rate is $S = 2 \times 0.32/(1 + 0.32) = 0.48$.

Genetic markers such as allozymes have been used widely to estimate F and S. The most commonly used model to estimate selfing and outcrossing is the mixed mating model. It assumes that there are only two types of matings, self-fertilization and random mating. However, matings among related individuals, such as full-sib, half-sib and cousin matings, also occur (**biparental inbreeding**). Consequently, the estimate of S is a measure of what the selfing rate would be, if all inbreeding was due to selfing.

Multilocus data provide more accurate estimates of true selfing rates. Further, in self-compatible plants, the difference between the mean of single locus estimates and the multilocus estimate provides an estimate of biparental inbreeding. For example, male-sterile individuals in seven populations of *Bidens* ssp. in Hawaii had average 'selfing' rates of 15%, and all of this must be due to biparental inbreeding. Similarly, the Pacific yew, the source of the anti-cancer compound taxol, is dioecious (separate sexes), but has an F of 47%. Again, all of this must be due to biparental inbreeding.

Parentage

Information on parentage is essential to study the impact of inbreeding, to verify pedigrees used in genetic management of threatened species, and to determine the effective size of populations (Chapter 4). Only rarely can parentage be determined from direct behavioural

Multiple DNA markers can be used to assign paternity and maternity observations. For example, female chimpanzees copulate with many males during their fertile periods. Genetic marker information from mother, offspring and putative fathers can be used to resolve these uncertainties.

If a paternally derived allele in the offspring is not present in the suspected father, then that male can be excluded as a potential father (unless a new mutation has occurred). If many loci are used, positive paternity assignments can be made with high probabilities. DNA fingerprints and multilocus microsatellites provide the most suitable markers. Figure 9.11 illustrates parentage determinations in snow geese, based on 14 nuclear RFLP loci. At several loci, the genotype of gosling 4 cannot be derived from those of its putative (candidate) parents by Mendelian inheritance. Possession of allele 3 at locus A excluded one or other putative parent as they are both 22 homozygotes. Similarly, gosling 4 possesses allele 2 at locus G and this allele is absent in both putative parents. Loci M and N exclude the putative father as he lacks the 1 allele. C excludes both parents as gosling 4 carries alleles not present in either putative parent.

Paternity determinations using microsatellites in captive chimpanzees revealed that the dominant male in the colony was responsible for siring most, but not all, of the offspring. Consequently, the need to move animals among zoos to minimize inbreeding and loss of genetic diversity is greater than if many males contributed to paternity, but it is less than if the dominant male fathered all offspring. As turtles mate at sea, and swimming individuals are very difficult to identify, their mating pattern is impossible to observe directly. Allozyme and microsatellite analyses of loggerhead turtle clutches established that females mated with several males.

Even in species where extensive behavioural observations have been made, genetic marker analyses have often revealed unexpected mating patterns. For example, splendid fairy wrens in Western Australia were reputed to have high rates of inbreeding, and no inbreeding depression. Subsequent paternity analyses using allozymes revealed that 65% or more of progeny were fathered by males from

RFLP locus

	A	B	C	D	E	F	G	H	I	J	K	L	M	N
Putative father	22	22	23	12	11	11	14	22	12	12	12	12	22	22
Putative mother	22	22	22	11	11	11	13	12	22	11	11	12	12	12
Gosling 1	22	22	22	12	11	11	11	12	12	11	12	11	12	12
Gosling 2	22	22	22	11	11	11	34	22	22	12	12	12	22	22
Gosling 3	22	22	22	12	11	11	13	22	12	11	11	22	22	22
Gosling 4	**23**	22	**11**	11	11	11	**12**	12	12	11	11	11	11	11

Fig. 9.11 Parentage determinations in snow geese. Genotypes at 14 nuclear RFLP loci are given for putative parents and goslings in a family of snow geese (after Avise 1994). Gosling 4 does not match its putative parents. Alleles that cannot be inherited from its putative parents are shown in bold and genotypes that cannot be derived from putative parents are underlined.

outside the group. Many birds with presumed monogamous mating systems have been shown to participate in extra-pair copulations. Even in humans, genetic markers have revealed that 10% or more of children are not the offspring of their registered father.

Pedigrees are used extensively in genetic management of captive populations and it is important to verify their accuracy. DNA fingerprinting in the critically endangered Waldrapp ibis identified five of 33 offspring whose pedigrees were incorrect and an additional unrelated founder. There also appear to be errors in the Bali starling, Arabian oryx, and Przewalski's horse studbooks.

The cases above may reflect a wider problem with inaccuracies in the pedigrees where errors may have serious implications in the management of endangered species.

Determining founder relationships

Small numbers of founders are frequently all that remain to initiate breeding programs of endangered species. Usually the relationships among founders are unknown. However, identification of related individuals is important for managing inbreeding and genetic diversity in the population. Genetic analyses using many loci (e.g. DNA fingerprinting or multiple microsatellite loci) can identify relationships in such cases. Studies in the California condor revealed three related groups of individuals amongst the 14 founders. Similar studies have been conducted on Bali starling, Guam rails, Micronesian kingfishers, Mauritius pink pigeons, Waldrapp ibises and Arabian oryx.

> Multilocus DNA markers can be used to delineate founder relationships

Sources of new founders

When numbers are small, all available founders should be used to establish captive breeding colonies. However, there may be uncertainties about the identity of some potential founders. These questions can be resolved using genetic analyses. For example, the Mexican wolf is extinct in the wild and the single Certified population was founded by only three or four animals. Two other populations existed, but it was unclear whether these had been subject to introgression from dogs, gray wolves or coyotes. Molecular genetic analyses (based on allozymes, mtDNA, DNA fingerprints and, particularly, microsatellites) established that all three populations of Mexican wolves were similar, with no detectable introgression (Fig. 9.12). The three populations are now being combined. The study also determined that the Certified population had three, not four founders. In a similar manner, two potentially new founders for the US captive population of Speke's gazelle were shown to be unrelated to US animals and have been added to the captive population.

> Where founder numbers are small, other potential founders can be examined, using genetic markers, to ensure that they belong to the correct species and are not affected by introgression

Sexing animals

Males and females of many bird and reptile species are morphologically indistinguishable. These must be sexed prior to pairing, as several cases of 'infertile' bird pairs in zoos have turned out to be two birds of the same sex.

> Birds and mammals can be sexed using genetic markers on the heterogametic sex chromosome

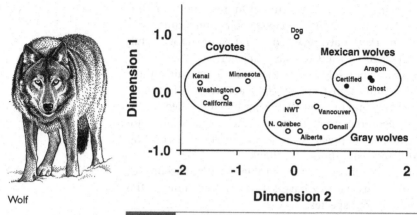

Wolf

Fig. 9.12 Are uncertified populations 'pure' Mexican wolves? Multidimensional scaling of allele frequency data from 20 microsatellite loci typed on populations of Mexican wolves, coyotes, dogs and gray wolves (Hedrick *et al.* 1997). The Certified population is known to consist of 'pure' Mexican wolves, while the Aragon and Ghost Ranch populations were of questionable status. Different coyote and gray wolf populations are indicated by state or province. *The Aragon and Ghost Ranch populations cluster with Certified as a group distinct from the other canids, indicating that all are 'pure' Mexican wolves.*

Some mammals are also difficult to sex, especially cetaceans (whales and dolphins) and secretive species, and it may not be possible to sex individuals when collecting samples by skin biopsies, hair, etc. The sex of stored DNA samples also may not be known.

Birds have ZZ male and ZW female sex chromosomes. Consequently, PCR primers for W chromosome specific sequences have been developed to distinguish males from females. W-specific fragments will amplify from the DNA of a ZW female, but not from a ZZ male (Fig. 9.13).

Birds can be sexed using genetic markers on the W sex chromosome

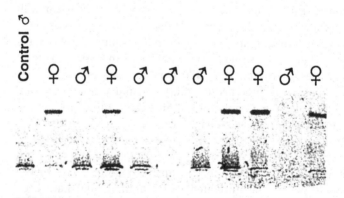

Fig. 9.13 Sexing the critically endangered Taita thrush using a W chromosome specific DNA marker (from Lens *et al.* 1998). PCR was used to amplify two loci, one on the W chromosome and one autosomal using specific primers. DNA amplified for all 11 birds, showing that the PCR reactions had worked. Results after digestion with the restriction enzyme Hae III are shown; this digests the autosomal fragment, but not the W fragment, so a fragment is found for ZW females, but not for ZZ males.

Molecular sexing is an important component in the program to recover the Norfolk Island boobook owl (Chapter 6). While the program had produced 12–13 individuals, of which seven were F_2, only two pairs were breeding. It was unclear whether this was due to hybrid sterility, unequal sex-ratio, or individuals of one sex being immature. As females and males could not be distinguished by external morphology, birds were sexed using a PCR-based technique. The population consisted of approximately equal numbers of females and males. A scarcity of mature males was the main factor slowing the recovery effort. The critically endangered black stilt in New Zealand had declined to about 70 birds with only 12 breeding pairs. Molecular sexing was used to avoid single-sex pairings in their recovery program. Molecular sexing in the critically endangered Taita thrush from Kenya (Fig. 9.13) has revealed a strongly distorted sex-ratio in one of three populations (only 10% females). Further, this study led to the identification of a morphological trait that can be used to sex individuals visually.

Since mammals typically have XX females and XY males, sex can be determined using molecular methods that detect Y chromosome specific loci. For example, free-ranging Pyrenean brown bears were sexed from hair and faecal samples found in the field using a PCR amplification of a Y-specific locus. Similarly, molecular methods have been developed to sex cetacean skin biopsy samples.

Mammals can be sexed using genetic markers on the Y chromosome

Disease

The disease status of animals is critical in identifying causes of population decline, and for checking candidates for translocation or reintroduction. PCR-based methods provide rapid, reliable and highly sensitive means for detecting disease organisms. For example, PCR has been used to study avian malaria in Hawaii, one of two diseases thought to have been major factors in the decline of Hawaiian birds. Higher-elevation habitats, considered free from malaria-carrying mosquitoes, have been preserved for endangered forest birds. However, malaria has been identified in blood of birds from high-elevation habitats on Maui and Hawaii, indicating that these areas are not as safe as previously thought. Reservoirs of the disease were also found in introduced bird species in low-elevation habitats.

Molecular methods provide means for detecting and investigating the biology of disease organisms, and delineating the source of new diseases

Gene trees based upon DNA sequences have been employed to determine the source of new diseases. HIV-1, one of the viruses that cause AIDS in humans, has been found to be most closely related to SIV virus from chimpanzees, while HIV-2 originated from a virus in sooty mangabeys. Similarly, an epidemic causing high mortality in African lions in the Serengeti in 1994 was shown to be due to canine distemper, presumed to have switched host species from local dogs. Recommendations were made to vaccinate local dogs against distemper to minimize the risk of repeat epidemics in lions and, especially, in other rarer carnivores. Molecular genetic analyses have also been used to identify the source of introduced plant diseases in Australia and elsewhere.

Diet

PCR with primers specific to suspected food items can be used to determine dietary items from gut contents or faeces

Bear

Diet is difficult to determine by direct observation in nocturnal and secretive species. Food items can be identified from faeces by using PCR with primers specific to suspected food items. This has been demonstrated in bears, where the plant *Photinia* was identified as a food item.

The role of predators in causing the decline of a threatened species has also been assessed using PCR-based amplification and genotyping. Microsatellite typing on stomach contents of glacous gulls in Alaska revealed that they were preying on emperor geese, but not on threatened spectacled eiders.

SUGGESTED FURTHER READING

Frankham, R., J. D. Ballou & D. A. Briscoe. 2002. *Introduction to Conservation Genetics*. Cambridge University Press, Cambridge, UK. Chapter 19 has a slightly extended treatments of these topics, plus references.

Avise, J. C. 1994. *Molecular Markers, Natural History and Evolution*. Chapman & Hall, New York. Has a lucid treatment of the use of molecular genetic techniques to understand species biology.

Avise, J. C. 2000. *Phylogeography: The History and Formation of Species*. Harvard University Press, Cambridge, MA. An excellent textbook on coalescence and phylogeography by the founder of the field.

Avise, J. C. & J. L. Hamrick. (eds.) 1996. *Conservation Genetics: Case Histories from Nature*. Chapman & Hall, New York. Scientific reviews containing examples of the use of genetic markers to understand species biology.

Hoelzel, A. R. (ed.) 1998. *Molecular Genetic Analysis: A Practical Approach*, 2nd edn. Oxford University Press, Oxford, UK. An edited book with technical details and case studies on the use of molecular genetic methods to illuminate aspects of species biology.

Smith, T. B. & R. K. Wayne. (eds.) 1996. *Molecular Genetic Approaches in Conservation*. Oxford University Press, New York. This book contains much technical detail and many examples of the use of molecular genetic analyses to understand species biology.

Sunnucks, P. 2000. Efficient markers for population biology. *Trends in Ecology and Evolution 15*, 199–203. Recent review on the use of molecular genetic methods in population biology and conservation.

Final messages

To obtain an overview of the book, we suggest you return to the **Take-home messages** and read them again. For students revising for examinations, you should find the boxes at the front of chapters plus the small main point boxes in the margin assist you in revision.

We trust that you have found this book informative, thought-provoking and interesting and that it will assist in your future conservation activities. The Earth's biodiversity is being lost at a frightening rate, so we must act now to conserve our life support system. Extinction is for ever.

Glossary

Adaptive evolution Changes in the genetic composition of populations due to natural selection that improve their reproductive fitness in a particular environment.

Additive variance The proportion of the genetic variance for a quantitative character in a population due to variation in the average effects of alleles. V_A is directly responsible for evolutionary potential and is a function of heterozygosity.

AFLP see *Amplified fragment length polymorphism*.

Allele An alternative form of a gene locus, e.g. wild-type vs. mutant, Fast vs. Slow electrophoretic mobility, copies of a microsatellite locus with different numbers of repeats of the AC sequence.

Allelic diversity A measure of genetic diversity within a population, computed as the average number of alleles per locus. For example, if the number of alleles at 6 loci are 1, 2, 3, 2, 1, and 1, then

$$\text{allelic diversity} = \frac{(1 + 2 + 3 + 2 + 1 + 1)}{6} = 1.67$$

Allopatric Populations, or species, whose geographic distributions do not overlap.

Allopolyploid A species whose chromosomal complement derives from two (or more) separate species (compare *Autopolyploid*); for example, tobacco with 48 chromosomes is an allotetraploid derived from two diploid species, each with 24 chromosomes. Many plant species have evolved in this manner. It is a form of instantaneous speciation.

Allozymes Alternative forms of a protein detected by electrophoresis that are due to alternative alleles at a single locus. Often referred to as isozymes.

Amino acid The building blocks of proteins.

Amplified DNA Many duplicated copies of a segment of DNA.

Amplified fragment length polymorphism (AFLP) A method for measuring genetic diversity. DNA is cut with a restriction enzyme, short synthetic DNA fragments of known sequence are added and the DNA amplified using PCR. The method produces a multilocus DNA fingerprint with polymorphism expressed as presence/absence of bands.

Assignment test Method of allocating individuals to their most likely groups based on multilocus microsatellite genotypes using clustering statistics. Used to detect migrants.

Autopolyploid A species derived by combining the two or more sets of chromosomes from the same species; for example, diploid populations of endangered grassland daisy have 22 chromosomes and autotetraploid populations have 44. Compare *Allopolyploid*.

Autoradiograpy Detection of radioactively labelled molecules by their effects in exposing photographic film.

Average heterozygosity A measure of genetic diversity within a population, computed as the sum of proportion heterozygous at all loci / total number of loci sampled. For example, if the heterozygosities at 5 loci are 0, 0.10, 0.20, 0.05 and 0, then

$$\text{average heterozygosity} = \frac{(0 + 0.10 + 0.20 + 0.05 + 0)}{5} = 0.07$$

Balancing selection Selection that maintains genetic variation in a population, encompassing heterozygote advantage (overdominance), frequency-dependent selection favouring rare genotypes, and forms of selection that vary over space, or time.

Binomial distribution A distribution describing the number of occurrences of two (or more) events in a sample of size n, e.g. the number of heads and tails in 50 tosses of a coin.

Biodiversity Biological diversity; the variety of ecosystems, species, populations within species and genetic diversity within living organisms.

Biological species concept This defines a species as a group of actually, or potentially, interbreeding individuals and natural populations that cannot interbreed with individuals from all other such groups, i.e. individuals within a species can exchange genetic material, while those from different species normally do not.

Bioresources Valuable products derived from the living world, including food, fibres and many drugs.

Biparental inbreeding Inbreeding due to production of offspring from mating of relatives more remote than self (full-sibs, half-sibs, cousins, etc.).

Bottleneck A restriction in population size.

Catastrophe An extreme event having a devastating impact on population or habitat e.g. cyclone, drought, extreme winter, disease epidemic.

Chloroplast DNA (cpDNA) Circular DNA molecules found in the chloroplasts of plants. They are usually maternally inherited.

CITES Convention on International Trade in Endangered Species.

Clade A sub-group of organisms from among a larger group sharing common ancestry, not shared by the other organisms in the larger group.

Cline Change in genetic composition of a population across a region or habitat gradient, such as a latitudinal cline, or an altitudinal cline. For example, the frequency of glaucous (waxy) leaves in several species of Eucalypt trees changes with altitude in several mountains in Tasmania, Australia. Clines may be due either to historical events (e.g. B blood group cline in humans) or to the balance between differential natural selection in different regions and migration between them (as for glaucous leaves).

Clones Individuals with identical genotypes, e.g. cuttings deriving from a single plant, or several individual animals derived from a single animal by nuclear transplantation.

Coalescence If two DNA sequence lineages converge at a common ancestral allele, they are said to coalesce.

Coancestry The coancestry of two individuals is the probability that two alleles, one from each individual, are identical by descent. Synonymous with kinship.

Common ancestor An individual that is an ancestor of both the mother and the father of a particular individual.

Corridor A ribbon of habitat connecting fragments that may allow migration.

cpDNA see *Chloroplast DNA*.

Critically endangered Description of a species with a very high probability of extinction within a short time, e.g. defined by IUCN as a 50% probability of extinction within 10 years, or three generations, whichever is longer.

Cryptic species Two or more species that are morphologically very similar, but show reproductive isolation, or clear genetic distinctiveness, e.g.

Chinese and Indian muntjac deer are morphologically similar, but have different chromosome numbers. Similar to *Sibling species*.

Demographic stochasticity Natural fluctuations in birth and death rates and sex-ratio. These may drive a small population to extinction. For example, the last six dusky seaside sparrows were, by chance, all males.

Diploid Having a pair of each chromosome.

Directional selection Selection in which the most extreme high (or low) individuals from a population are parents of the next generation.

DNA fingerprint The 'barcode' produced by probing for minisatellite repeat sequences on the DNA of an individual.

Dominance Deviation of heterozygote phenotype from the mean of homozygotes at a locus, say in the direction of the wild-type homozygote compared to the mutant homozygote. Compare *Recessive*.

Dominance variance The proportion of quantitative genetic variation for a character in a population due to the deviations of heterozygotes from the average effects of homozygotes.

Ecological exchangeability Concept used in defining management units within species; also termed ecological replaceability. If two populations are adapted to similar environments then they are exchangeable.

Ecosystem services Essential functions supplied free of charge by living organisms, including nutrient recycling, pest control, and pollination of crop plants.

Effective population size (N_e) The number of individuals that would result in the same inbreeding or genetic drift if they behaved in the manner of an idealized population.

Effectively neutral The situation where the selective forces on an allele are so weak that it behaves as if it is not subject to natural selection. Occurs when the selection coefficient is less than $1/(2N_e)$, where N_e is the effective population size.

Electrophoresis A method for separating proteins or DNA fragments in a gel according to their net charges, shape and size.

Endangered Description of a species or populations with a high probability of extinction within a short time, defined by IUCN as a 20% probability within 20 years or 10 generations, whichever is longer.

Environmental stochasticity Natural fluctuations in environmental conditions, such as rainfall, food supply, competitors, winter temperatures, etc. These may drive a small population to extinction.

Environmental variance That portion of the phenotypic variation for a particular quantitative character due to environmental sources, e.g. the variation exhibited by a completely homozygous population.

ESU see *Evolutionarily significant unit*.

Evolution Change in the genetic composition of a population.

Evolutionary potential The ability of a population to evolve to cope with environmental changes, such as those due to climate change or changed disease organisms. Usually equated with genetic diversity, as this is required for evolution to occur, but reproductive rate and population size also contribute to it.

Evolutionarily significant units (ESU) Partially genetically differentiated populations that are considered to require management as separate units.

Exchangeability Concept used in defining management units within species; also termed replaceability. If the genetic compositions of two populations are similar, they are exchangeable. If two populations are adapted to similar environments then they are exchangeable.

Expected heterozygosity The heterozygosity expected for a random mating population with the given allele frequencies according to the Hardy–Weinberg equilibrium. For example, if the allelic frequencies at a locus are 0.2 and 0.8, the expected heterozygosity is $2pq = 2 \times 0.2 \times 0.8 = 0.32$.

Ex situ Away from its normal habitat, such as an endangered species being conserved in captivity, or an endangered plant being preserved in a seed store or cryopreserved.

Extinction Permanent disappearance of a population or species.

Extinction vortex The likely interaction between human impacts, inbreeding and demographic fluctuation in a downward spiral towards extinction.

F Wright's inbreeding coefficient; fixation index.

F **statistics** Measures of total inbreeding in a population (F_{IT}), partitioned into that due to inbreeding within sub-populations (F_{IS}) and that due to differentiation among sub-populations (F_{ST}).

F_{IS} That proportion of the total inbreeding within a population due to inbreeding within sub-populations.

F_{IT} The total inbreeding in a population, due to both inbreeding within sub-populations (F_{IS}) and differentiation among sub-populations (F_{ST}).

F_{ST} The proportion of the total inbreeding in a population due to differentiation among sub-populations.

Fitness Reproductive fitness, i.e. the number of fertile offspring surviving to reproductive age contributed by an individual.

Fixation The situation where all individuals in a population are identically homozygous, e.g. all $A_1 A_1$.

Fixation index Wright's inbreeding coefficient *F*.

Forensics Detection of illegal activities by scientific means. For example, DNA-based methods are being developed to tests for tiger material in Asian medicines following amplification of DNA using PCR.

Founder effect Change in the genetic composition of a population due to origin from a small sample of individuals; a single-generation bottleneck. Founder effects usually result in loss of genetic diversity, extinction of alleles, genetic drift and an increase in inbreeding.

Full-sibs Individuals sharing the same two parents. A full-sib mating is between a brother and a sister.

Gene diversity See *Expected heterozygosity*.

Gene flow Movement of alleles between populations due to migration.

Gene tree Tree showing the relationships between different copies of a single locus.

Genetic distance A measure of the genetic difference between allele frequencies in two populations, or species. The most commonly used form is Nei's genetic distance.

Genetic diversity The extent of genetic variation in a population or species, or across a group of species, e.g. heterozygosity, or allelic diversity, or heritability.

Genetic drift Changes in the genetic composition of a population due to random sampling in small populations. It results in loss of genetic diversity, random changes in allele frequencies and diversification among replicate populations. Also referred to as *Random genetic drift*.

Genetic exchangeability Concept used in defining management units within species; also termed genetic replaceability. If the genetic compositions of two populations are similar, they are exchangeable.

Genetic load The load of deleterious alleles in a population, some due to the balance between deleterious mutation and selection (mutation load) and some due to heterozygote advantage and other forms of balancing selection (balanced load).

Genetic marker A locus or karyotype that provides useful information about distinctiveness of populations or taxa, or information useful in forensics, etc. Information may come from allozymes, microsatellites, chromosomes, or inherited morphological characters.

Genetic stochasticity Genetic effects in small populations that have a chance element, including inbreeding, loss of genetic diversity and mutational accumulation that may drive a population or species to extinction.

Genome The complete genetic material of a species, or individual; i.e., all of the DNA, all of the loci, or all of the chromosomes.

Genome resource bank A store containing a diversity of genetic material for one or more species, including seed stores, cryopreserved gametes, embryos, somatic cells, or a collection of DNA samples.

Genotype × environment interaction Differential performance of genotypes in different environments. For example, many plants have races (populations) that grow and survive better in their home environment than in other environments.

Haplotype Haploid genotype; the allelic composition for several different loci on a chromosome, e.g. $A_1B_3C_2$.

Haplotype network A diagram showing different haplotypes joined by lines to show relationships, typically with a line joining each haplotype differing by a single base in DNA sequence.

Hardy–Weinberg equilibrium The equilibrium genotype frequencies achieved in a random mating population with no perturbing forces from mutation, migration, selection or chance. If two alleles A_1 and A_2 have frequencies of p and q, the Hardy–Weinberg equilibrium frequencies for the A_1A_1, A_1A_2 and A_2A_2 genotypes are p^2, $2pq$ and q^2, respectively.

Harmonic mean Reciprocal of the arithmetic mean of reciprocals

$$\frac{1}{H} = \frac{1}{n} \sum_{i=1}^{n} \frac{1}{X_i}$$

where H is the harmonic mean, n the sample size and X_i the ith observation. The harmonic mean of population sizes in different generations describes the impact of population size fluctuations on the effective population size.

Heritability Proportion of the variation for a quantitative character due to genetic causes.

Hermaphrodite An animal or plant with both sexes present in single individuals; this condition is found in many plants and some animals, such as snails. Also known as monoecious.

Heterozygosity Proportion of heterozygotes in a population, most often measured over several gene loci. See *Average heterozygosity, Expected heterozygosity, Observed heterozygosity.*

Heterozygote An individual with two different forms of a gene at a locus, e.g. A_1A_2.

Heterozygote advantage A form of selection where the heterozygote has a higher fitness than the homozygotes (overdominance), e.g. sickle-cell anaemia in humans in malarial areas. This results in active maintenance of genetic variation in very large populations. It is one form of balancing selection.

Homozygote An individual with two copies of the same allele at a gene locus, e.g. $A_1 A_1$.

Hybrid zone An area of overlap where different sub-species or species mix and breed, producing hybrid offspring.

Idealized population A conceptual random mating population with equal numbers of hermaphrodite individuals breeding in each generation, no mutation, migration, or selection, and Poisson variation in family sizes (mean = variance = 1). Used as a standard to which other populations are equated when defining effective population sizes.

Identity by descent Alleles that are identical copies of an allele in a common ancestor.

Inbreeding Production of offspring from the mating of individuals related by descent, e.g. self-fertilization, brother–sister, or cousins matings.

Inbreeding coefficient (F) The most commonly used measure of the extent of inbreeding; the probability that two alleles at a locus in an individual are identical by descent. It has a range from 0 to 1.

Inbreeding depression Reduction in reproduction, survival or other quantitative characters due to inbreeding.

In situ conservation Conservation of a species in its natural wild habitat.

Interaction variance The proportion of the quantitative genetic variation for a character in a population due to the deviation of genotypic effects from the average effects of the constituent loci.

Introgression Introduction of genetic material from another species or sub-species into a population. It is a threat to the genetic integrity of a range of canid, fish, plant, etc species.

Intron A region of a locus that is transcribed into mRNA, but not translated.

IUCN The World Conservation Union. The initials originally stood for the International Union for Conservation of Nature, later expanded to included 'and Nature Reserves', but the organization now refers to itself by the first-mentioned name.

Kinship (k_{ij}) The kinship of two individuals is the probability that two alleles, one from each individual, are identical by descent; synonymous with *Coancestry*. Equivalent to the inbreeding coefficient of an offspring of the two individuals, if they had one.

Lethal Inconsistent with survival, as in a recessive lethal allele that results in death when homozygous.

Lethal equivalents (B) A measure for comparing the extent of inbreeding depression in different populations. A group of detrimental alleles that would cause death if homozygous, e.g. one lethal equivalent equals one lethal allele, two alleles each with a 50% probability of causing death, etc. It is determined as the slope of the regression of natural logarithm of survival on the inbreeding coefficient F.

Lineage sorting Chance fixation of different alleles (or haplotypes) in different lineages (populations or species) when the initial population is polymorphic; it may lead to incorrect phylogenies.

Locus (plural loci) A segment of DNA at a particular location on a chromosome. Often called a gene locus.

Major histocompatibility complex (MHC) A large family of loci that play an important role in the vertebrate immune system. They produce molecules that bind foreign antigens. These loci show extraordinarily high levels of genetic diversity.

Management unit A population within a species that is sufficiently genetically differentiated from other populations to warrant separate management.

Mean kinship The average kinship of an individual with all other individuals in a population, including itself. Minimizing kinship is the current recommended method for genetically managing endangered species in captivity.

Meta-analysis A statistical analysis that uses the combined information from several different studies, or several different species.

Metapopulation A group of partially isolated populations of the same species that undergo local extinction and recolonizations.

MHC See *Major histocompatibility complex.*

Microsatellite A locus with a short tandem repeat DNA sequence, such as the AC sequence repeated 10 times. Microsatellites typically show variable number of repeats and high heterozygosities in populations.

Minimizing kinship The current recommended method for genetically managing endangered species in captivity. It maximizes the retention of genetic diversity.

Minimum viable population size (MVP) The minimum size of population that will be viable in the long term, meaning a probability of extinction of say 1% in 1000 years. The initial sizes were derived from genetic considerations, but it rapidly became clear that demographic and environmental stochasticity and catastrophes must be considered as well.

Minisatellite A region of DNA, usually in the 10s to 100s of bases in length, that shows variation in number of repeats; also known as variable number tandem repeats (VNTR). When several such loci are probed they result in a DNA fingerprint that looks like a barcode.

Mismatch analysis Analysis based on the distribution of number of DNA sequence differences among pairs of alleles.

Mitochondrial DNA (mtDNA) The circular DNA molecule contained within mitochondria; typically maternally inherited.

Monomorphic The presence of only one allele at a locus in a population, generally taken to mean the most common allele is at a frequency of greater than 99%, or 95%; lack of genetic diversity. Contrast with *Polymorphic.*

mtDNA See *Mitochondrial DNA.*

Mutation A sudden genetic change, i.e. parents lack the condition, but it occurs in one or more offspring and is heritable.

Mutation load The load of deleterious mutations carried in a population. Homozygosity for such mutations is considered to be the main cause of inbreeding depression.

Mutation–selection balance The equilibrium due to the occurrence of deleterious mutation and the forces of natural selection removing them, resulting in low frequencies of deleterious mutations (mutation load).

'Mutational meltdown' The decline in reproductive rate and downward spiral towards extinction due to chance fixation of new mildly deleterious mutations in small populations.

MVP see *Minimum viable population size.*

Natural selection Mortality or altered reproductive rates due to natural environmental processes. Differential reproduction and survival of genotypes (natural selection) leads to adaptive evolutionary change.

Nei's genetic distance (D_N) The most widely used measure of the genetic difference between allele frequencies in two populations, or species, devised by Masatoshi Nei. It is calculated as the natural logarithm of *Nei's genetic similarity* (I_N).

Nei's genetic similarity (I_N) The most widely used measure of the genetic similarity between allele frequencies in two populations, or species, devised by Masatoshi Nei.

Neutral mutation A mutation that is equivalent in effects on reproductive fitness to the existing allele, such that its fate is determined by chance effects associated with population size (*Random genetic drift*).

Neutral theory Predictions and theory based on sampling for neutral alleles.

Normal distribution A symmetrical bell-shaped distribution with a characteristic mean and variance. Many quantitative characters show approximately normal distributions.

Observed heterozygosity The actual level of heterozygosity measured in a population, i.e. if there are two alleles at a locus, F and S, and a sample of individuals contain 60 FF, 30 FS and 10 SS, the observed heterozygosity is 30%. It is typically averaged across several loci, say 20–50 allozyme loci.

Outbreeding Not inbreeding; a population that is not undergoing deliberate inbreeding. Approximately the same as *Random mating*.

Outbreeding depression A reduction in reproductive fitness due to crossing of two populations (or sub-species, or species).

Parthogenesis Reproduction without fertilization.

PCR see *Polymerase chain reaction*.

Pedigree A chart specifying lines of descent and relationship.

Percentage of loci polymorphic (P) A measure of genetic diversity within a population, computed as (number of polymorphic loci/total number of loci sampled) $\times$ 100. For example if 3 loci are polymorphic, and 7 are monomorphic,

$$P = \frac{3}{10} \times 100 = 30\%.$$

Phenotypic variance The variance in phenotype for a quantitative character in a particular population.

Phylogenetic tree A tree reflecting the relationships between different species or populations.

Phylogeny The evolutionary development and history of a sub-species, species or higher taxa. Often visualized as a *Phylogenetic tree*.

Phylogeography The field of study concerned with the geographical distribution of genealogical lineages, especially within species. Typically DNA sequence trees are related to the geographical origins of haplotypes.

Poisson distribution A statistical distribution with variance equal to the mean. It is used to predict the number of occurrences of rare events, such as the distribution of families of sizes 0, 1, 2, 3, . . . and is assumed to occur in the *Idealized population*.

Polygamy Mating with more than one individual of the opposite sex.

Polymerase chain reaction (PCR) Method used to amplify specific segments of DNA. The DNA is heated, primers (short segments of DNA flanking the segment of interest) added and the intervening DNA copied using thermostable *Taq* polymerase enzyme. Usually 30–40 cycles of amplification are performed in a thermocycler, each consisting of separation of complementary DNA strands at 94 °C, annealing of primers at 50–60 °C depending on primer sequences, and extension (copying) at 72 °C.

Polymorphic The presence of more than one allele at a locus in a population. Generally taken to mean the most common allele is at a frequency of less than 99%, or 95%; the existence of genetic diversity. Compare *Monomorphic*.

Polyploid Having more than two copies of the haploid genetic complement, e.g. $4n$ is tetraploid.

Population viability analysis (PVA) The process of predicting the fate of a population (including risk of extinction) due to the combined effects of all deterministic and stochastic threats faced by a population. PVA is also used as a management tool to examine the impacts of different management options to recover threatened species. Typically done by inputting life history information into computer software.

Primer A short nucleotide sequence that pairs with one strand of DNA and provides a free end at which DNA polymerase enzyme begins synthesis of a complementary segment of DNA.

Probe DNA from a known locus used to hybridize with other DNA via complementary base pairing to identify similar sequences in the other DNA. The probe is usually radioactively labelled (e.g. with ^{32}P) so that fragments of DNA showing homology are detected using autoradiography.

Probed Process where DNA from an individual is hybridized with labelled DNA from a known locus.

Purging Elimination of deleterious alleles from populations due to natural selection, especially that associated with populations subject to inbreeding.

PVA See *Population viability analysis*.

QTL See *Quantitative trait locus*.

Quantitative character Characters such as size, reproductive rate, survival, etc. that typically have continuous approximately normal distributions. Variation for them is controlled by many loci and by environmental conditions.

Quantitative genetic variation Genetic variation affecting a quantitative character, such as size, reproductive rate, behaviour or chemical composition. Presumed to be due to the cumulative effects at many loci (QTL).

Quantitative trait locus (QTL) A locus affecting a quantitative character.

Random amplified polymorphic DNA (RAPD) Genetic diversity detected following PCR amplification using random primers of DNA (usually 10 bases in length) to amplify random segments of DNA. Polymorphisms are detected as presence vs. absence of bands.

Random genetic drift Changes in the genetic composition of a population due to random sampling in small populations. It results in loss of genetic diversity, random changes in allele frequencies and diversification among replicate populations; frequently referred to as genetic drift.

Random mating A pattern of mating (also termed random breeding) where the chances of two genotypes or phenotypes breeding is determined by their frequencies in the population, e.g. if AA and aa have frequencies of

Variance The most commonly used measure of dispersion among quantitative measurements; the square of the standard deviation. The average of the squared deviation from the mean, computed as

$$V = \sum_{i=1,}^{n} \frac{(X_i - \bar{X})^2}{(n-1)}$$

where X_i is the ith observation, $\bar{X}$ is the mean, and n is the total number of observations.

VNTR Variable number tandem repeat. Also referred to as DNA fingerprint and minisatellites.

Vulnerable A species or population with a tangible risk of extinction within a moderate time, defined by IUCN as a 10% probability within 100 years.

Sources and copyright acknowledgments

We are grateful to the following for kind permission to reproduce copyright material:

Chapter 3 frontispieces: Oxford University Press: from Plate 8.2 in Kettlewell, H. B. D. (1973) *The Evolution of Melanism*, Clarendon Press, Oxford, UK.

Box 3.3 map: Blackwell Publishing Ltd: from Fig. 1 in Kettlewell, H. B. D. (1958) A survey of the frequencies of *Biston betularia* (L.) (Lep.) and its melanic forms in Great Britain. *Heredity 12*, 551–572.

Fig. 3.2: © Oxford University Press, 1976. Reprinted from Mourant, A. E., C. Kopé & K. Domaniezska-Sobczak (1976) *The Distribution of Human Blood Groups and Other Polymorphisms*, 2nd edn. Oxford University Press, Oxford, UK.

Fig. 3.4: The Carnegie Institute of Washington: from Figs. 19, 23 and 25 in Clausen, J., D. D. Keck & W. M. Heisey (1940) *Experimental Studies on the Nature of Species*, vol. 1. Carnegie Institute of Washington Publication no. 520, Washington, DC.

Box 4.2 map: Center for Applied Studies in Forestry, Steven F. Austin State University: from James, F. (1995) The status of the red-cockaded woodpecker and the prospects for recovery. In Kulhavy, D. L., P. G. Hooper & R. Costa (eds.) *The Red-Cockaded Woodpecker: Recovery, Ecology and Management*. Center for Applied Studies, Steven F. Austin State University, Nacogdoches, TX.

Fig. 4.4: MIT Press: from Fig. 2 in Foose, T. J. (1986) Riders of the last ark. In Kaufman, L. & K. Mallory (eds.) *The Last Extinction*. MIT Press, Cambridge, MA.

Fig. 6.2: CSIRO Publishing: from Johnston, P. G., R. J. Davey & J. H. Seebeck (1984) Chromosome homologies in *Potoroos tridactylis* and *P. longipes* based on G-banding patterns. *Australian Journal of Zoology 32*, 319–324.

Fig. 7.3: Blackwell Publishing Ltd: from Zhi, L., W. E. Johnson, M. A. Menotti-Raymond, N. Yuhki, J. S. Martenson, S. Mainka, H. Shi-Quiang, Z. Zhihe, G. Li, W. Pan, X. Mao & S. J. O'Brien (2001) Patterns of genetic diversity in remaining giant panda populations. *Conservation Biology 15*, 1596–1607.

Fig. 9.9: Copyright (1995) National Academy of Sciences: from Bowen, B. W. F., A. Abreu-Grobius, G. H. Balazas, N. Kamenzaki, C. J. Limpus & R. J. Ferl (1995) Trans-Pacific migrations of the loggerhead turtle (*Caretta caretta*) demonstrated with mitochondrial DNA markers. *Proceedings of the National Academy of Sciences, USA 92*, 3731–3734.

Fig. 9.10: Oxford University Press: from Fritch, P. & L. H. Reiseberg (1996) The use of random amplified DNA in conservation genetics. In *Molecular Genetic Approaches to Conservation*, eds. Thomas B. Smith & Robert K. Wayne, copyright 1996 by Oxford University Press, Inc. Used by permission of Oxford University Press, Inc.

Fig. 9.12: John Wiley & Sons, Inc.: from Hedrick, P. W., P. S. Miller, E. Greffen & R. Wayne (1997) Genetic evaluation of the three captive Mexican wolf lineages. *Zoo Biology 16*, 47–69. Reprinted with permission of Wiley-Liss, Inc., a subsidiary of John Wiley & Sons, Inc.

Fig. 9.13: Kluwer Academic Publishers: from Lens, L., P. Galbusera, T. Brooks, E. Waiyaki & T. Schenck (1998) Highly skewed sex ratios in the critically endangered Taitra thrush as revealed by CHD genes. *Biodiversity and Conservation* 7, 869–873.

SOURCES FOR FIGURES AND TABLES

Avise, J. C. 1994. *Molecular Markers, Natural History, and Evolution.* Chapman & Hall, New York.

Avise, J. C. 2000. *Phylogeography: The History and Formation of Species.* Harvard University Press, Cambridge, MA.

Baker, C. S. & S. R. Palumbi. 1996. Population structure, molecular systematics, and forensic identification of whales and dolphins. Pp. 10–49 in J. C. Avise & J. L. Hamrick, eds. *Conservation Genetics: Case Histories from Nature.* Chapman & Hall, New York.

Ballou, J. D. & R. C. Lacy. 1995. Identifying genetically important individuals for management of genetic diversity in pedigreed populations. Pp. 76–111 in J. D. Ballou, M. Gilpin & T. J. Foose, eds. *Population Management for Survival and Recovery: Analytical Methods and Strategies in Small Population Conservation.* Columbia University Press, New York.

Ballou, J. D., R. C. Lacy, D. Kleiman, A. Rylands & S. Ellis. 1998. *Leontopithecus II: The Second Population and Habitat Viability Assessment for Lion Tamarins (Leontopithecus): Final Report.* Conservation Breeding Specialist Group (SSC/IUCN), Apple Valley, MN.

Barone, M. A., M. E. Roelke, J. Howard, J. L. Brown, A. E. Anderson & D. E. Wildt. 1994. Reproductive characteristics of male Florida panthers: comparative studies from Florida, Texas, Colorado, Latin America, and North American Zoos. *J. Mammal.* 75, 150–162.

Beheregaray, L. B., P. Sunnucks, D. L. Alpers, S. C. Banks & A. C. Taylor. 2000. A set of microsatellite loci for the hairy-nosed wombats (*Lasiorhinus krefftii* and *L. latifrons*). *Conserv. Genet.* 1, 89–92.

Beheregaray, L. B., C. Ciofi, A. Caccone, J. P. Gibbs & J. R. Powell. 2002. Genetic divergence, phylogeography and conservation units of giant tortoises from Santa Cruz and Pinzón, Galápagos Islands. *Conserv. Gen.* 4, 31–46.

Berger, J. 1990. Persistence of different sized populations: an empirical assessment of rapid extinctions in bighorn sheep. *Conserv. Biol.* 4, 91–98.

Bowen, B. W., F. A. Abreu-Grobois, G. H. Balazs, N. Kamenzaki, C. J. Limpus & R. J. Ferl. 1995. Trans-Pacific migrations of the loggerhead turtle (*Caretta caretta*) demonstrated with mitochondrial DNA markers. *Proc. Natl. Acad. Sci. USA* 92, 3731–3734.

Brook, B. W. & J. Kikkawa. 1998. Examining threats faced by island birds: a population viability analysis on the Capricorn silvereye using long-term data. *J. Appl. Ecol.* 35, 491–503.

Clausen, J., D. D. Keck & W. M. Heisey. 1940. *Experimental Studies on the Nature of Species*, vol. 1, *Environmental Responses of Climatic Races of Achillea.* Carnegie Institute, Washington, DC.

Constable, J. L., M. V. Ashley, J. Goodall & A. E. Pusey. 2001. Noninvasive paternity assignment in Gombe chimpanzees. *Molecular Ecology* 10, 1279–1300.

Crandall, K. A., O. R. P. Bininda-Edmonds, G. M. Mace & R. K. Wayne. 2000. Considering evolutionary processes in conservation biology: an alternative to 'evolutionary significant units'. *Trends Ecol. Evol.* 15, 290–295.

Culver, M., W. E. Johnson, J. Pecon-Slattery & S. J. O'Brien. 2000. Genomic ancestry of the American puma (*Puma concolor*). *J. Hered. 91*, 186–197.

Daniels, S. J., J. A. Priddy & J. R. Walters. 2000. Inbreeding in small populations of red-cockaded woodpeckers: analyses using a spatially-explicit simulation model. Pp. 129–147 in A. G. Young & G. M. Clarke, eds. *Genetics, Demography and Viability in Fragmented Populations*. Cambridge University Press, Cambridge, UK.

Daugherty, C. H., A. Cree, J. M. Hay & M. B. Thompson. 1990. Neglected taxonomy and continuing extinctions of tuatara (*Sphenodon*). *Nature 347*, 177–179.

De Bois, H., A. A. Dhondt & B. Van Puijenbroek. 1990. Effects of inbreeding on juvenile survival of the okapi *Okapi johnstoni* in captivity. *Biol. Conserv. 54*, 147–155.

Dizon, A., G. Lento, S. Baker, P. Parsboll, F. Capriano & R. Reeves. 2000. *Molecular genetic identification of whales, dolphins, and porpoises: Proceedings of a Workshop on the forensic Use of Molecular Techniques to Identify Wildlife Products in the Marketplace: NOA Technical Memorandum NMFS*. US Department of Commerce, Washington, DC.

Dobson, A. P., G. M. Mace, J. Poole & R. A. Brett. 1992. Conservation biology: the ecology and genetics of endangered species. Pp. 405–430 in R. J. Berry, T. J. Crawford & G. M. Hewitt, eds. *Genes in Ecology*. Blackwell, Oxford, UK.

El-Kassaby, Y. A. & A. D. Yanchuk. 1994. Genetic diversity, differentiation, and inbreeding in Pacific yew from British Columbia. *J. Hered. 85*, 112–117.

Eldridge, M. D. B., J. M. King, A.K. Loupis, P. B. S. Spencer, A. C. Taylor, L. C. Pope & G. P. Hall. 1999. Unprecedented low levels of genetic variation and inbreeding depression in an island population of the black-footed rock wallaby. *Conserv. Biol. 13*, 531–541.

England, P. R. 1997. Conservation genetics of population bottlenecks. PhD thesis, Macquarie University, Sydney, NSW, Australia.

Fenner, F. & F. N. Ratcliffe. 1965. *Myxomatosis*. Cambridge University Press, Cambridge, UK.

Frankham, R. 1995. Inbreeding and extinction: a threshold effect. *Conserv. Biol. 9*, 792–799.

Fritsch, P. & L. H. Rieseberg. 1996. The use of random amplified polymorphic DNA (RAPD) in conservation genetics. Pp. 54–73 in T. B. Smith & R. K. Wayne, eds. *Molecular Approaches in Conservation*. Oxford University Press, New York.

Futuyma, D. J. 1998. *Evolutionary Biology*, 3rd edn. Sinauer, Sunderland, MA.

Gottelli, D., C. Sillero-Zubiri, G. D. Appelbaum, M. S. Roy, D. J. Girman, J. Garica-Moreno, E. A. Ostrander & R. K. Wayne. 1994. Molecular genetics of the most endangered canid: the Ethiopian wolf *Canis simensis*. *Mol. Ecol. 3*, 301–312.

Grant, B. S. 1999. Fine tuning the peppered moth paradigm. *Evolution 53*, 980–984.

Grativol, A. D., J. D. Ballou & R. C. Fleischer. 2001. Microsatellite variation within and among recently fragmented populations of the golden lion tamarin (*Leontopithecus rosalia*). *Conserv. Genet. 2*, 1–9.

Groombridge, J. J., C. G. Jones, M. W. Bruford & R. A. Nichols. 2000. 'Ghost' alleles of the Mauritius kestrel. *Nature 403*, 616.

Hamrick, J. L. & M. J. W. Godt. 1989. Allozyme diversity in plant species. Pp. 43–63 in A. H. D. Brown, M. T. Clegg, A. L. Kahler & B. S. Weir, eds.

Plant Population Genetics, Breeding, and Genetic Resources. Sinauer, Sunderland, MA.

Hedrick, P. W. 1983. *Genetics of Populations*. Science Books International, Boston, MA.

Hedrick, P. W. 2000. *Genetics of Populations*, 2nd edn. Jones & Bartlett, Sudbury, MA.

Hedrick, P. W., P. S. Miller, E. Geffen & R. Wayne. 1997. Genetic evaluation of the three Mexican wolf lineages. *Zoo Biol. 16*, 47–69.

Houlden, B. A., P. R. England, A. C. Taylor, W. D. Greville & W. B. Sherwin. 1996. Low genetic variability of the koala *Phascolarctos cinereus* in southeastern Australia. *Mol. Ecol. 5*, 269–281.

Houlden, B. A., B. H. Costello, D. Sharkely, E. V. Fowler, A. Melzer, W. Ellis, F. Carrick, P. R. Baverstock & M. S. Elphinstone. 1998. Phylogeographic differentiation in the mitochondrial control region in the koala, *Phascolarctos cinereus* (Goldfuss 1817). *Mol. Ecol. 8*, 999–1011.

Houle, D., B. Morikawa & M. Lynch. 1996. Comparing mutational variabilities. *Genetics 143*, 1467–1483.

IUCN. 1996. *1996 IUCN Red List of Threatened Animals*. IUCN, Gland, Switzerland.

Kettlewell, H. B. D. 1973. *The Evolution of Melanism*. Clarendon Press, Oxford, UK.

Kulhavy, D. L., R. G. Hooper & R. Costa. 1995. *Red-cockaded Woodpecker: Recovery, Ecology and Management*. Center for Applied Studies, Stephen F. Austin State University, Nacogdoches, TX.

Land, D. E. & R. C. Lacy. 2000. Introgression level achieved through Florida panther genetic restoration. *Endang. Sp. Updates 17*, 99–103.

Lens, L., P. Galbusera, T. Brooks, E. Waiyaki & T. Schenck. 1998. Highly skewed sex ratios in the critically endangered Taita thrush as revealed by CHD genes. *Biodiver. Conserv. 7*, 869–873.

Lu, Z., W. E. Johnson, M. A. Menotti-Raymond, N. Yuhki, J. S. Martenson, S. Mainka, H. Shi-Quiang, Z. Zhihe, G. Li, W. Pan, X. Mao & S. J. O'Brien. 2001. Patterns of genetic diversity in remaining giant panda populations. *Conserv. Biol. 15*, 1596–1607.

Lynch, M. & B. Walsh. 1998. *Genetics and Analysis of Quantitative Traits*. Sinauer, Sunderland, MA.

Madsen, T., R. Shine, M. Olsson & H. Wittzell. 1999. Restoration of an inbred adder population. *Nature 402*, 34–35.

Majerus, M. E. N. 1998. *Melanism: Evolution in Action*. Oxford University Press, Oxford, UK.

Milne, H. & F. W. Robertson. 1965. Polymorphism in egg albumen and behaviour in eider ducks. *Nature 205*, 367–369.

Moritz, C., J. Worthington Wilmer, L. Pope, W. B. Sherwin, A. C. Taylor & C. J. Limpus. 1996. Applications of genetics to the conservation and management of Australian fauna: four case studies from Queensland. Pp. 442–456 in T. B. Smith & R. K. Wayne, eds. *Molecular Genetic Approaches to Conservation*. Oxford University Press, New York.

Norman, J., P. Olsen & L. Christidis. 1998. Molecular genetics confirms taxonomic affinities of the endangered Norfolk Island boobook *Ninox novaeseelandiae undulata*. *Biol. Conserv. 86*, 33–36.

Nunney, L. & K. A. Campbell. 1993. Assessing minimum viable population size: demography meets population genetics. *Trends Ecol. Evol. 8*, 234–239.

Pitman, N. C. A. & P. M. Jørgensen. 2002. Estimating the size of the World's threatened flora. *Science 298*, 989.

Princée, F. P. G. 1995. Overcoming the constraints of social structure and incomplete pedigree data through low-intensity genetic management. Pp. 124–154 in J. D. Ballou, M. Gilpin & T. J. Foose, eds. *Population Management for Survival and Recovery: Analytical Methods and Strategies in Small Population Conservation*. Columbia University Press, New York.

Ralls, K. & J. Ballou. 1983. Extinction: lessons from zoos. Pp. 164–184 in C. M. Schonewald-Cox, S. M. Chambers, B. MacBryde & L. Thomas, eds. *Genetics and Conservation: A Reference for Managing Wild Animal and Plant Populations*. Benjamin/Cummings, Menlo Park, CA.

Ralls, K., J. D. Ballou, B. A. Rideout & R. Frankham. 2000. Genetic management of chondrodystrophy in the California condor. *Anim. Conserv. 3*, 145–153.

Reed, D. H., J. J. O'Grady, B. W. Brook, J. D. Ballou & R. Frankham. 2003. Estimates of minimum viable population size for vertebrates and factors influencing those estimates. *Biol. Conserv. 113*, 23–34.

Rich, S. S., A. E. Bell & S. P. Wilson. 1979. Genetic drift in small populations of *Tribolium*. *Evolution 33*, 579–583.

Rieseberg, L. H. & S. M. Swensen. 1996. Conservation genetics of endangered island plants. Pp. 305–334 in J. C. Avise & J. L. Hamrick, eds. *Conservation Genetics: Case Histories from Nature*. Chapman & Hall, New York.

Roelke, M. E., J. Martenson & S. J. O'Brien. 1993. The consequences of demographic reduction and genetic depletion in the endangered Florida panther. *Curr. Biol. 3*, 340–350.

Saccheri, I., M. Kuussaari, M. Kankare, P. Vikman, W. Fortelius & I. Hanski. 1998. Inbreeding and extinction in a butterfly metapopulation. *Nature 392*, 491–494.

Seebeck, J. H. & P. G. Johnson. 1980. *Potorous longipes* (Marsupialia: Macropidae): a new species from Eastern Victoria. *Aust. J. Zool. 28*, 119–134.

Seymour, A. M., M. E. Montgomery, B. H. Costello, S. Ihle, G. Johnsson, B. St John, D. Taggart & B. A. Houlden. 2001. High effective inbreeding coefficients correlate with morphological abnormalities in populations of South Australian koalas (*Phascolarctos cinereus*). *Anim. Conserv. 4*, 211–219.

Sherwin, W. B., P. Timms, J. Wilcken & B. Houlden. 2000. Analysis and conservation implications of koala genetics. *Conserv. Biol. 14*, 639–649.

Sloane, M. A., P. Sunnucks, D. Alpers, L. B. Beheregaray & A. C. Taylor. 2000. Highly reliable genetic identification of individual northern hairy-nosed wombats from single remotely collected hairs: a feasible censusing method. *Mol. Ecol. 9*, 1233–1240.

Smith, H. G. 1993. Heritability of tarsus length in cross-fostered broods of the European starling (*Sturnus vulgaris*). *Heredity 71*, 318–322.

Smith, T. B., L. A. Freed, J. K. Lepson & J. H. Carothers. 1995. Evolutionary consequences of extinctions in populations of a Hawaiian honeycreeper. *Conserv. Biol. 9*, 107–113.

Soulé, M. E. & D. Simberloff. 1986. What do genetics and ecology tell us about the design of nature reserves? *Biol. Conserv. 35*, 19–40.

Stangel, P. W., M. R. Lennartz & M. H. Smith. 1992. Genetic variation and population structure of Red-cockaded woodpeckers. *Conserv. Biol. 6*, 283–292.

Sukumar, R., U. Ramakrishnan & J. A. Santosh. 1998. Impact of poaching on an Asian elephant population in Periyar, southern India: a model of demography and tusk harvest. *Anim. Conserv. 1*, 281–291.

Sun, M. 1996. The allopolyploid origin of *Spiranthes hongkongensis* (Orchidaceae). *Am. J. Bot. 83*, 252–260.

Tarr, C. L., S. Conant & R. C. Fleischer. 1998. Founder events and variation at microsatellite loci in an insular passerine bird, the Laysan finch (*Telespiza cantans*). *Mol. Ecol. 7*, 729–731.

Taylor, A. C., W. B. Sherwin & R. K. Wayne. 1994. Genetic variation of microsatellite loci in a bottlenecked species: the northern hairy-nosed wombat *Lasiorhinus krefftii*. *Mol. Ecol. 3*, 277–290.

Taylor, A. C., A. Horsup, C. N. Johnson, P. Sunnucks & W. B. Sherwin. 1997. Relatedness structure detected by microsatellite analysis and attempted pedigree reconstruction in an endangered marsupial, the northern hairy-nosed wombat *Lasiorhinus krefftii*. *Mol. Ecol. 6*, 9–19.

Templeton, A. R. 1998. Nested clade analyses of phylogeographic data: testing hypotheses about gene flow and population history. *Mol. Ecol. 7*, 381–397.

Thomas, C. D. 1990. What do real populations tell us about minimum viable population sizes? *Conserv. Biol. 4*, 324–327.

Ward, R. D., D. O. F. Skibinski & M. Woodwark. 1992. Protein heterozygosity, protein structure, and taxonomic differentiation. *Evol. Biol. 26*, 73–159.

Weigensberg, I. & D. A. Roff. 1996. Natural heritabilities: can they be reliably estimated in the laboratory? *Evolution 50*, 2149–2157.

Westemeier, R. L., J. D. Brawn, S. A. Simpson, T. L. Esker, R. W. Jansen, J. W. Walk, E. L. Kershner, J. L. Bouzat & K. N. Paige. 1998. Tracking the long-term decline and recovery of an isolated population. *Science 282*, 1695–1698.

Xu, X. & U. Arnason. 1996. The mitochondrial DNA molecule of Sumatran orangutan and a molecular proposal for two (Bornean and Sumatran) species of orangutan. *J. Mol. Evol. 43*, 431–437.

Zhi, L., W. B. Karesh, D. N. Janczewski, H. Frazier-Taylor, D. Sajuthi, F. Gombek, M. Andau, J. S. Martenson & S. J. O'Brien. 1996. Genetic differentiation among natural populations of orangutan (*Pongo pygmaeus*). *Curr. Biol. 6*, 1326–1336.

Index

Page numbers in bold type refer to figures; those in italics refer to material in tables.